"十三五"江苏省高等学校重点教材
（编号：2017-2-164）

大学专门用途英语系列教材

English for Environmental Science and Engineering
环境科学与工程英语

主　审／杨金才
总主编／肖　飞
主　编／祝　平
副主编／赵　诚
编　者／曹顺娣
　　　　陈振娇
　　　　巩丽娜
　　　　孙新征
　　　　王　琤
　　　　张知国
　　　　刘　芳
　　　　刘　宏
　　　　李　丽

外语教学与研究出版社
FOREIGN LANGUAGE TEACHING AND RESEARCH PRESS
北京 BEIJING

图书在版编目 (CIP) 数据

环境科学与工程英语 / 祝平主编；曹顺娣等编. －－ 北京：外语教学与研究出版社，2020.12 (2024.5 重印)
大学专门用途英语系列教材 / 肖飞总主编
ISBN 978-7-5213-2250-7

Ⅰ. ①环… Ⅱ. ①祝… ②曹… Ⅲ. ①环境科学－英语－高等学校－教材②环境工程－英语－高等学校－教材 Ⅳ. ①X-43

中国版本图书馆 CIP 数据核字 (2020) 第 257202 号

出 版 人　王　芳
责任编辑　牛亚敏
责任校对　胡春玲
版式设计　袁　凌
封面设计　锋尚设计
出版发行　外语教学与研究出版社
社　　址　北京市西三环北路 19 号（100089）
网　　址　https://www.fltrp.com
印　　刷　河北虎彩印刷有限公司
开　　本　787×1092　1/16
印　　张　11
版　　次　2020 年 12 月第 1 版 2024 年 5 月第 4 次印刷
书　　号　ISBN 978-7-5213-2250-7
定　　价　49.90 元

如有图书采购需求，图书内容或印刷装订等问题，侵权、盗版书籍等线索，请拨打以下电话或关注官方服务号：
客服电话：400 898 7008
官方服务号：微信搜索并关注公众号"外研社官方服务号"
外研社购书网址：https://fltrp.tmall.com

物料号：322500001

前言

根据《大学英语教学指南》的精神，大学英语的课程体系主要由通用英语、专门用途英语和跨文化交际三大类课程组成。

大学专门用途英语系列教材充分体现《大学英语教学指南》的精神，在大学英语教学改革实践的基础上，以培养与专业英语相关的英语能力为目标，将特定的学科内容与英语语言学习相结合，兼顾语言输入与输出训练，帮助学生实现在英语语境下对学科知识的有效输出和应用。

大学专门用途英语系列教材依据以内容为依托的教学理念编写，具有时代感、知识性和实用性。教材所选内容反映学科主线，体现相关学科的基本知识和前沿信息，兼具专业性和可读性。基于课文内容设计的阅读理解、专业词汇和学术英语词汇练习，帮助学生在理解课文的同时掌握文章中重要词汇，同时注重活学活用和适度扩展。此外，教材还提供设计灵活、注重实效的思辨训练和学术技能训练，帮助学生在实践中提高思辨能力、习得学术规范、培养学术研究能力，从而能够有效、得体地使用英语进行学业学习与学术交流。

大学专门用途英语系列教材能满足学生专业发展的需要，同时保证他们在大学期间的英语语言水平稳步提高。丰富的教学内容和多样的练习形式也为实现分类教学和因材施教提供可能，教师可根据实际需要选择教学内容，制定个性化的教学方案。

大学专门用途英语系列教材的编者们恳请使用者对本书中出现的问题提出宝贵意见和建议，以便再版时改进。

大学专门用途英语系列教材编委会

2017.6

Contents

Unit 1
Environmental science and engineering P1

Text A	Studying environmental science: definition, scope and importance P3
Text B	What is environmental engineering and what do environmental engineers do? P18

Unit 2
Air pollution P23

Text A	Air pollution: causes and solutions P25
Text B	How air pollution works on different scales P38

Unit 3
Water pollution P43

Text A	Water pollution: an introduction P45
Text B	Solutions to water pollution P58

Unit 4
Soil pollution P63

Text A	What is soil pollution? P65
Text B	Reducing soil pollution and erosion P80

Unit 5
Solid waste and disposal P85

Text A	Solid waste disposal in the U.S. P87
Text B	Better solid waste disposal P103

Unit 6
Waste recycling P107

Text A	Recycle more? Or…recycle better? P109
Text B	Single stream recycling P125

Unit 7
Ecosystems P129

Text A	Major types of ecosystems P131
Text B	What is ecology? P144

Unit 8
Environmental protection P149

Text A	Principles and approaches of environmental protection P151
Text B	Environmental degradation: Causes and effects P166

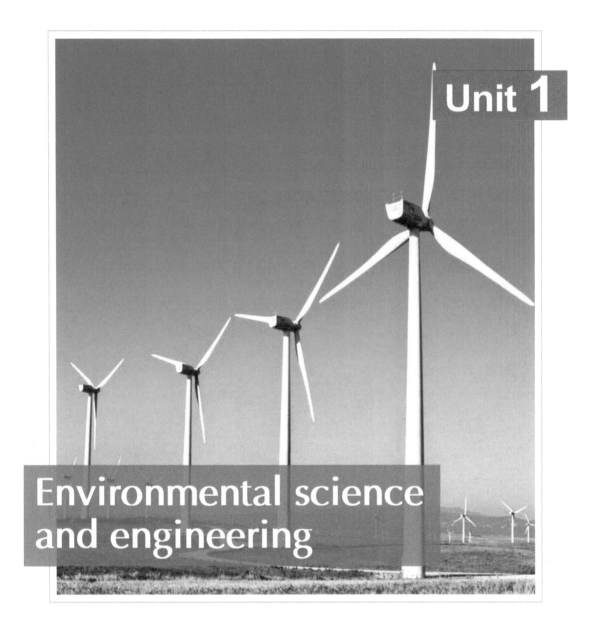

Unit 1

Environmental science and engineering

In this unit, you will learn:

- **Subject-related knowledge:** The concept of environmental science
  The importance of environmental engineering
- **Academic skill:** Searching for information
- **Reading strategy:** Dealing with unknown words (Part I)

Section A

Pre-reading

1 Match the expressions with the pictures below.

air purification   water reclamation   garbage sorting
soil conservation   urban landscaping   waste recycling

1. _____

2. _____

3. _____

4. _____

5. _____

6. _____

2 Discuss the following questions with your partner.

1. What do you know about environmental science?
2. What kinds of environmental problems are human beings facing?

Studying environmental science: definition, scope and importance

Text A

1. Environmental science is the academic field that takes physical, biological and chemical sciences to study the environment and discover solutions to environmental problems. Environmental science is an interdisciplinary subject, which can be a challenge as the fields it covers require different skills and knowledge. However, by combining an understanding of all of these areas, students are better able to study the environment from an integrated perspective. Environmental science also branches out into environmental studies and environmental engineering. In a world where global warming, air pollution, and plastic waste are major topical issues, environmental science is becoming an increasingly valued and relevant discipline. Environmental science combines elements of the key traditional fields of chemistry, biology, etc. Many people have questions about what studying environmental science is like and where it can lead. This text will answer these questions and a number of other common questions about environmental science.

2. **What do you study in environmental science?**
Environmental science involves different fields of study. Most often, the study of environmental science includes the study of climate change, natural resources, energy, pollution, and other environmental issues. In environmental sciences, ecologists study how organisms interact with each other, chemists study chemical processes occurring in the environment, geologists study the formation, structure and history of earth, biologists study the biodiversity, and physicists study physical phenomenon, rules, etc. in the environment.

3 The growing complexity of environmental problems are creating a need for scientists with rigorous, interdisciplinary training in environmental science. Environmental scientists and specialists use their knowledge of the natural sciences to protect the environment and human health. They must also have a solid background in economics, sociology and political science.

4 **Importance of Environmental Science**
The importance of environmental science lies as follows:

5 • **To realize that environmental problems are global**
Environmental science lets you recognize that environmental problems such as climate change, global warming, ozone layer depletion, acid rains, and impacts on biodiversity and marine life are not just national problems, but global problems as well. So, concerted effort from across the world is needed to tackle these problems.

6 • **To understand the impacts of development on environment**
It's well documented and quantified that development results in industrial growth, urbanization, expansion of telecommunication and transport systems, hi-tech agriculture and expansion of housing. Environmental science seeks to teach the general population to highlight the interrelationship between social-economic development and environment. The goal is to achieve development sustainably without compromising the future generation's ability to satisfy their own needs.

7 • **To discover sustainable ways of living**
Environmental science is more concerned with discovering ways to live more sustainably. This means utilizing present resources in a manner that conserves their supplies for the future. Environmental sustainability doesn't have to outlaw living luxuriously, but it advocates creating awareness about consumption of resources and minimizing unnecessary waste. This includes minimizing household energy consumption, using disposals to dispose of

waste, eating locally, recycling more, growing your own food, conserving household water, and driving your car less.

8 • **To utilize natural resources efficiently**
Natural resources bring a whole lot of benefits to a country. A country's natural resources may not be utilized efficiently because of low-level training and lack of management skills. Environmental science teaches us how to use natural resources efficiently, for example, how to appropriately put into practice environmental conservation methods and how to use the right tools to explore resources.

9 • **To shed light on contemporary concepts of biodiversity conservation**
Biodiversity is the variety of life on earth. The present rate of biodiversity loss is at an all-time high. Environmental science aims to teach people how to reverse this trend. For example, we are advised to use sustainable wood products, eco-friendly cleaning products, and consume sustainable seafood. We can also support or take active part in the conversation campaigns at local levels.

10 • **To understand the interrelationship between organisms and humans**
Organisms and humans depend on each other to get by. Environmental science is important because it enables you to understand how these relationships work. For example, humans breathe out carbon dioxide, which plants need for photosynthesis. Plants, on the other hand, produce and release oxygen to the atmosphere, which humans need for respiration. Animal droppings are sources of nutrients for plants and other microorganisms. Plants are sources of food for humans and animals. In short, organisms and humans depend on each other for survival.

11 • **To learn and create awareness about environmental problems at local, national and international levels**
Environmental problems at local, national and international levels mostly

occur due to lack of awareness. Environmental science aims to educate and equip learners with necessary environmental skills to pass to the community in order to create awareness. Environmental awareness can be created through social media, community-centered green clubs, and forums.

New words and expressions

academic /ˌækəˈdemɪk/ *adj.* 学术的

interdisciplinary /ˌɪntəˈdɪsɪplɪnəri/ *adj.* drawing from or characterized by participation of two or more fields of study 学科间的；跨学科的；涉及多个学科的

integrated /ˈɪntɪɡreɪtɪd/ *adj.* formed or united into a whole 综合的；整合的；融合的

discipline /ˈdɪsɪplɪn/ *n.* a branch of knowledge 学科

biodiversity /ˌbaɪəʊdaɪˈvɜːsəti/ *n.* the existence of a large number of different kinds of animals and plants which make a balanced environment 生物多样性

rigorous /ˈrɪɡərəs/ *adj.* demanding that particular rules, processes, etc. are strictly followed 严格的

depletion /dɪˈpliːʃən/ *n.* the act of decreasing sth. markedly 消耗；损耗

marine /məˈriːn/ *adj.* connected with the sea and the creatures and plants that live there 海洋的；海生的

concerted /kənˈsɜːtɪd/ *adj.* done in a planned and determined way, especially by more than one person, government, or country, etc. 同心协力的；努力的

document /ˈdɒkjumənt/ *vt.* to record the details of sth. 记录，记载

quantify /ˈkwɒntɪfaɪ/ *vt.* to describe or express sth. as an amount or a number 量化

urbanization /ˌɜːbənaɪˈzeɪʃən/ *n.* the process of creating cities or towns in country areas 城市化

compromise /ˈkɒmprəmaɪz/ *vt.* to harm or damage sth. in some way, for example, by behaving in a way that does not match a legal or moral standard 损害，危害，危及

sustainable /səˈsteɪnəbl/ *adj.* involving the use of natural products and energy in a way that does not harm the environment 可持续的

outlaw /ˈaʊtlɔː/ *vt.* to make sth. illegal 宣布…不合法；使…成为非法

advocate /ˈædvəkeɪt/ *vt.* to support sth. publicly 拥护；支持；提倡

disposal /dɪˈspəʊzəl/ *n.* a kitchen appliance for getting rid of garbage 垃圾处理器

photosynthesis /ˌfəʊtəʊˈsɪnθɪsɪs/ *n.* the process by which green plants turn carbon dioxide and water into food using energy obtained from sunlight 光合作用

respiration /ˌrespəˈreɪʃən/ *n.* the act of breathing 呼吸

nutrient /ˈnjuːtriənt/ *n.* a substance that is needed to keep a living thing alive and to help it grow 营养物；营养素

microorganism /ˌmaɪkrəʊˈɔːɡənɪzəm/ *n.* a very small living thing that you can only see through a microscope 微生物

forum /ˈfɔːrəm/ *n.* a place where people can exchange opinions; a meeting organized for this purpose 论坛；讨论会

ozone /ˈəʊzəʊn/ **layer** 臭氧层

carbon dioxide /ˈkɑːbən daɪˈɒksaɪd/ *n.* 二氧化碳

acid rain 酸雨

dispose /dɪˈspəʊz/ **of** 处理；除掉；解决

shed light on 阐明；解释

Reading comprehension

1 Read Text A and complete the chart to get the outline and main idea of the text.

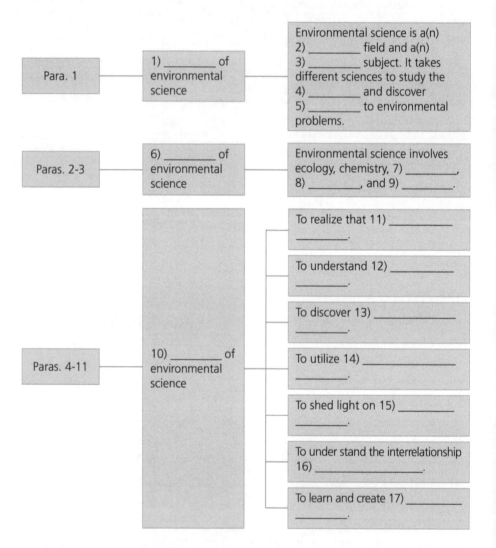

Para. 1	1) _____ of environmental science	Environmental science is a(n) 2) _____ field and a(n) 3) _____ subject. It takes different sciences to study the 4) _____ and discover 5) _____ to environmental problems.
Paras. 2-3	6) _____ of environmental science	Environmental science involves ecology, chemistry, 7) _____, 8) _____, and 9) _____.
Paras. 4-11	10) _____ of environmental science	To realize that 11) _____.
		To understand 12) _____.
		To discover 13) _____.
		To utilize 14) _____.
		To shed light on 15) _____.
		To under stand the interrelationship 16) _____.
		To learn and create 17) _____.

2 Read the text and decide whether the following statements are true or false. If false, underline the mistakes and put the corrections in the blanks provided.

1. Environmental science is an interdisciplinary subject which involves studying elements of many natural sciences instead of social sciences.
   _____

2. Laboratory work is also a key part of studying environmental science. _____

3. The efforts of environmental scientists are enough for solving environmental problems. _____

4. Environmental sustainability has to outlaw living luxuriously. _____

5. A country's natural resources may not be utilized efficiently because of low-level training and lack of management skills. _____

6. The present rate of biodiversity increase is at an all-time high. _____

7. Environmental science is important because it enables you to understand how natural organisms work. _____

8. Environmental problems at local, national and international levels mostly occur due to lack of natural resources. _____

Language focus

1 Match the words and phrases in Column A with the definitions in Column B and give their Chinese meanings in Column C.

Column A	Column B	Column C
___ 1. marine	a. to use something for a particular purpose	_____
___ 2. sustainable	b. the way that green plants make their food using sunlight	_____
___ 3. disposal	c. an organism of microscopic or ultramicroscopic size	_____
___ 4. utilize	d. a kitchen appliance that cuts garbage into small pieces	_____
___ 5. biodiversity	e. the existence of different species of plants and animals	_____
___ 6. urbanization	f. relating to the sea or to the animals and plants living in the sea	_____
___ 7. photosynthesis	g. able to continue without causing damage to the environment	_____
___ 8. microorganism	h. the process of becoming or growing into a city	_____

2 Fill in the blanks with the words given below. Change the form where necessary.

ozone   organism   urban   marine
dispose   respiration   sustainable   nutrient

1. Nowadays many cities and towns have grown rapidly as people have moved from the countryside, which is called _____ and creates increasing pressure on environment.

2. It is recognized that global climate change has been the most urgent ecological issue facing the planet, so everyone has a responsibility to live more simply and _____ in order to reduce carbon impacts.

3. Oceans, seas and other _____ resources, which help regulate the global ecosystem, are essential to human well-being and economic development worldwide, especially for people living in coastal communities.

4. A lot of fish suffered severe damage from lack of oxygen in the water because of excessive _____ from industrial waste and sewage waste.

5. In the forum, representatives from the UNICEF said that many children have got _____ diseases because of air pollution.

6. This factory has invested more than one billion dollars to accelerate its development of technologies used for the _____ and recycling of wastes.

7. According to a recent study, excessive exhaust of waste gas into the air has caused global warming and _____ layer depletion.

8. Farmers in this area produce _____ agricultural products without using artificial chemicals, which is market-friendly and eco-friendly.

3 Translate the following paragraph into English.

环境科学与工程是一个跨学科专业，不仅涉及生物学、地理学、生态学、化学、物理等自然学科，还涉及到经济学、社会学、政治学等领域。目前，从全球范围来看，环境问题十分严峻，因此，该学科的重要性不言而喻。不仅科学家和研究人员需要深入了解相关知识，普通民众也应增强环境意识，学会有效地利用资源。各国政府应该联合起来，同心协力地解决环境问题，以实现环境的可持续发展。

Critical thinking

1. Look at the following pictures and form groups to debate the two different modes of social development.

2. Do you think that everything we use is going to be 100% recyclable in the future? Work in groups to discuss what life will be like in the future with the rapid development of environmental science. Then each group gives a short report to the class.

Research task

Academic skill: Searching for information

Information can come from virtually anywhere — personal experiences, media, blogs, books, journal and magazine articles, expert opinions, encyclopedias, and web pages, etc.

1. Types of information

    Please note that only the typical uses are listed in the table for each type of source. The uses of each type of source may overlap.

Type	Typical use
Magazine	• To find information or opinions about popular culture • To find up-to-date information about current events • To find non-scholarly articles about topics of interest within the subject of the magazine
Academic journal	• To get help for your scholarly research • To find out what has been studied on your topic • To find bibliographies that point to other relevant research
Database	• To find articles on specific topics • To find online journals or news articles
Newspaper	• To find editorials, commentaries, expert or popular opinions • To find current local, national or world news
Library catalog	• To find virtually any topic • To find hard copies of current or back issue of journals, books, newspapers or magazines
Website	• To find public information from all levels of government – from central to local • To find expert or popular opinions • To find information of various types of media, e.g. illustrations, audio and video

2. Searching for information

    Searching by author / title

    Searching by author / title obviously assumes that you are searching for a particular author, book or article, probably in either a database or a library catalog. Here are some tips:

    - When searching by author, put the author's last name first, e.g. "Kotler, Philip", not "Philip Kotler", if he is from an English-speaking country. Search the author's full name in Chinese order if he is Chinese. Sometimes, the author could be an organization, so give the full name of the organization as it commonly appears, e.g. "World Bank".

- When searching by title, enter the correct title as accurately as possible.

Searching by keyword

It is basically a way of searching through subject or topic. Most library catalogs and databases include an option to search by keyword as an alternative to author and title. The first step of keyword search is to decide the key word(s) or phrase(s). Normally, the word(s) or phrase(s) which can cover or indicate the topic you search for can be selected as keyword(s). A good research topic usually contains two or three concepts. For example, you need to write a paper on "The Impact of Cognitive Styles on Design Students' Spatial Knowledge". We can break the topic into concepts, like "cognitive styles" and "spatial knowledge", which can be used as keywords. Then type them in a search bar in a database, for instance, EBSCOhost. In a database, there are usually two ways of search, i.e. basic search and advanced search.

Basic search (see Fig. 1) generates a large number of sources for you to differentiate, which is an exhausting task. But advanced search (see Fig. 2), which provides more choices for further conditioning, can make the work lighter. There are many variables that can be chosen to refine the search. And you can define the relationship between the keywords by choosing "and", "or" or "not" based on the results you intend to obtain.

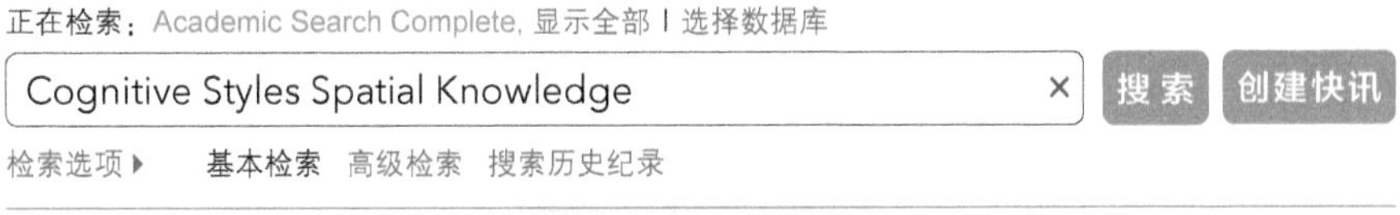

Fig. 1 Basic search

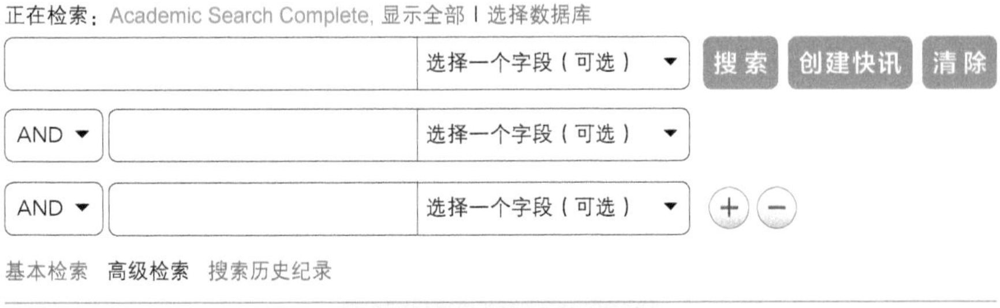

Fig. 2 Advanced search

As "cognitive styles" is a broader topic and "spatial knowledge" is more specific, they can be typed in the upper and middle search bars respectively. More relevant results will appear. You can then refine the search by selecting a specific variable. In this case, "SU 主题语" can be chosen to filter the results (See Fig. 3).

Snowball search

正在检索：Academic Search Complete, 显示全部 | 选择数据库

Cognitive Styles		SU 主题语 ▼	搜索 创建快讯 清除
AND ▼	Spatial Knowledge	选择一个字段（可选）▼	
AND ▼		选择一个字段（可选）▼	⊕ ⊖

基本检索 高级检索 搜索历史纪录

精确搜索结果	检索结果：1-9（共9个）
当前检索 ▼	
布尔逻辑词组： SU cognitive styles AND spatial knowledge	1. The Impact of Cognitive Styles on Design Students' Spatial Knowledge

Fig. 3

It is a good way if your topic has a keyword or author. You can trace the citations of that author using a specialized citation database, such as the Social Science Citation Index to obtain other key works or authors. You will follow the stream of research up to the near present and see the way in which the work or the author has influenced the subsequent studies.

3. Evaluating information

Once you have found information that satisfies the requirements of your research, you should evaluate it. Evaluating information encourages you to think critically about the reliability, validity, accuracy, authority, timeliness, point of view or bias of information.

When evaluating information, you can use the five criteria AAOCC, namely, Authority, Accuracy, Objectivity, Currency and Coverage. They can be applied to check all information.

1) Authority of information
   - Who published it?

- What institution published it?
- Does the publisher list his or her qualifications?
- Who provided it, and can you contact him or her?

2) Accuracy of information
   - Does it provide enough details?
   - Has it been cited correctly?

3) Objectivity of information
   - What is the purpose of it, or why was it published?
   - Is it biased?
   - What opinions (if any) are expressed by the author?

4) Currency of information
   - When was it published?
   - When was it updated?
   - How up-to-date is it?

5) Coverage of information
   - Do citations in it complement the research?
   - Is it all text or a balance of text and image?

Task

Now you know what environmental science is and what an environmental scientist does. Work in groups and search some information on the general knowledge that an environmental scientist should acquire. Evaluate the information using the AAOCC criteria. Then write down what you have found, where and how you have found them and share them in groups.

Section B

Reading strategy

Dealing with unknown words (Part I)

The ability to deal with unknown words is a key reading skill in the reading process. It is a vital skill because you are almost certain to find unknown or unfamiliar words in any text. The skill is not necessarily to "know" the words, but to guess the meaning of them so that you can read and understand the whole text. Here are several ways that can help you guess the meaning of an unknown word.

Guessing by explanation

Sometimes, you will find that the meaning of an unfamiliar word is given to you in the text. In this case, what you need to do is keep reading when you find an unfamiliar word and keep on reading. Typically, you can get the meaning from a phrase immediately after the unfamiliar word. For example,

> **To shed light on contemporary concepts of biodiversity conservation.**
> Biodiversity is the variety of life on earth. (Para.9)

From the definition we can figure out that biodiversity refers to the different life forms on earth.

Guessing by synonyms

This is an excellent skill to learn. What you do here is looking at other words which relate to that word and work out what it may mean. These words are called synonyms (words with a similar meaning). For example,

> However, by combining an understanding of all of these areas, students are better able to study the environment from an integrated perspective. (Para.1)

In this example, you may have no idea of the meaning of "integrate", but you can infer the meaning from the previous word "combine" in "by combining an understanding of all of these areas". Here "combine" can well be a synonym of the later word "integrate".

Guessing by common sense and experience

Sometimes, when you come across an unknown word, you can guess the meaning of it by your common sense. For example,

> For example, humans breathe out carbon dioxide, which plants need for photosynthesis. Plants, on the other hand, produce and release oxygen to the atmosphere, which humans need for respiration. (Para.10)

In the sentence "humans breathe out carbon dioxide, which plants need for photosynthesis", we can guess the meaning of the word "photosynthesis" through the description. Because the description includes a common sense about how plants use carbon dioxide, water and light to survive. Thus we can figure out the meaning of "photosynthesis".

Task

Read Text B and apply the skills above to deal with the underlined words.

Text B

What is environmental engineering and what do environmental engineers do?

1. Environmental engineering is the branch of engineering that is concerned with protecting people from the effects of adverse environmental effects, such as pollution, as well as improving environmental quality. A closely-related field in environmental studies is environmental engineering. Environmental engineering is the application of science and engineering principles to help improve the Earth's environment to provide healthier land, water and air for human use and to find ways to scale back on pollution sites. It uses the principles of biology and chemistry to develop solutions to environmental problems. The goal of environmental engineering is to integrate scientific and engineering principles to minimize air pollution and waste discharge from industries, clean up polluted sites, set up appropriate mechanism for waste disposal resulting from human activity and study the impact of proposed construction sites on the environment.

2. The practice of environmental engineering dates back to the dawn of civilization. Ever since groups of people began living in semi-permanent settlements, they have had to deal with the challenges of providing clean water and disposing of solid waste and sewage. With the growth of cities and the advent of large-scale farming and manufacturing, people have also had to worry about air quality and soil contamination.

discharge *n.* 排放

3 The first environmental engineer is said to have been Joseph Bazalgette. According to an article in the *Postgraduate Medical Journal*, Bazalgette oversaw the construction of the first large-scale municipal sanitary sewer system in London in the mid-19th century. This was prompted by a series of cholera epidemics, as well as a persistent unbearable stench, that were attributed to the discharge of raw sewage into the Thames River, which was also the main source of drinking water for the city. This "great stink", which was so noxious that it caused Parliament to evacuate Westminster, gave then Prime Minister Benjamin Disraeli grounds to ask for 3.5 million pounds to improve the city's sewage disposal system.

4 Apart from this, environmental engineering promotes energy conservation and is also concerned with finding solutions in areas of public health such as implementing adequate sanitation facilities.

5 Topics involving environmental engineering include waste management, water supply, waste water treatment, air pollution, recycling, disposal of wastes, protection from radiation, public health, environmental engineering law, and the effects of man-made projects on the environment.

6 Briefly speaking, the main task of environmental engineering is to protect public health by protecting (from further degradation), preserving (the present condition), and enhancing the environment. There are various divisions of the field of environmental engineering.

7 **Environment impact and mitigation.** In this division, engineers and scientists apply scientific and engineering principles to evaluate if there is likely to be an adverse impact on water, air, soil, land quality, flora and fauna, or any ecological impact, noise impact or visual impact. The overall goal of environmental engineers is to identify, assess and evaluate the environmental impacts of man-made plans, projects and laws to see if there are any affects

---

cholera *n.* 霍乱

whatsoever on water, air, habitat, agriculture, plants and animals, and the ecosystem. In this sense, environmental engineering encompasses the protection of human health and preserving the natural environment using scientific and engineering processes.

8 **Solid waste management.** Solid waste management deals with solid waste materials produced by human activity, whether straightforwardly or obliquely. Solid waste management also focuses on the recovery of resources to delay the human consumption of natural resources. Different methods are used to limit the harmful solids that are released into the environment. The objective of solid waste management is to reduce the harmful effects of solid, liquid or gaseous substances on the environment.

9 Solid waste management also encompasses reuse and recycling of everyday materials, the minimization of waste, the storage and transfer of wastes, disposing of solid wastes at landfills, and policies and regulations regarding the above.

10 **Water supply and treatment.** In this area, environmental engineers work on civilian and agricultural use of water. It is their job to assess water within a water basin and determine supply of available water, the cycles of water movement throughout the seasons, and the treatment of water for various uses. This is done to severely minimize the risk of diseases that can be caught from the drinking and / or contact of water. This is why water distribution systems are built to meet irrigation standards for civilian use.

11 **Water pollution.** Water pollution is closely tied to water supply and treatment. It essentially deals with the waste put into water from sewer systems, outhouses and septics in both rural and urban areas. Environmental engineers design specific systems to carry the polluted water away from civilized areas and release the waste into the environment, most often into the ocean, and, in more landlocked areas, rivers and lakes.

12 **Air pollution management.** Here, environmental engineers design manufacturing and combustion processes to ensure the air pollutants are at acceptable levels in the atmosphere where the toxins can have minimal effects on people. These include emissions from automobile exhausts, NO_2, SO_2, organic compounds, and organic acids.

13 Apart from social and ecological impact, environmental engineering is also involved in protection of wildlife. Environmental engineers also work as consultants and provide their services to clients to protect our environment from various environmental hazards, and clean up hazardous sites. They might also work in such areas as environmental policy and regulation development and so on.

14 In the field of environmental engineering, engineers take doctrines of engineering, science, biology and chemistry to formulate solutions to public health issues and environmental problems. Essentially, they study the effects technology has on the environment. Environmental engineers also work to improve recycling, waste disposal, public health, and water and air pollution control.

15 One of the most important responsibilities of environmental engineers is to prevent the release of harmful chemical and biological contaminants into the air, water and soil. This requires extensive knowledge of the chemistry and biology of the potential contaminants as well as the industrial or agricultural processes that might lead to their release. With this knowledge, new processes can be designed, or existing processes can be modified, to reduce or eliminate the release of pollutants.

16 Another important function performed by environmental engineers is detecting the presence of pollutants and tracking them back to their source. In some cases, this can present a significant challenge. For instance, the source of contamination in a lake could be anywhere within several thousands of acres

of land surrounding the lake and its tributaries. Contamination of oceans can present even greater challenges in identifying the source.

17 Once the environmental engineer identifies a source of contamination, it must be stopped or significantly reduced. Simply shutting down a business is not always a viable option, because of the potential for severe economic consequences. Environmental engineers often work with businesses to determine ways to avoid or reduce the production of pollutants or to separate them so they can be disposed of in a safe manner.

18 Critical skills needed by environmental engineers include a working knowledge of chemical engineering, fluid dynamics, geography, geology and hydrology. Also, because of the numerous legal issues involved and the prevalence of litigation in environmental issues, environmental engineers must be familiar with applicable laws, and many of them are also practicing attorneys.

Unit 2

Air pollution

In this unit, you will learn:

- **Subject-related knowledge:** Causes of and solutions to air pollution
  Different scales of air pollution
- **Academic skill:** Collecting data
- **Reading strategy:** Dealing with unknown words (Part II)

Section A

Pre-reading

1 Match the expressions with the pictures below.

global warming   greenhouse effect   forest fire
factory emission   automobile exhaust   ozone layer

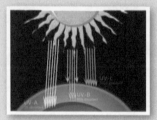
1. _____

2. _____

3. _____

4. _____

5. _____

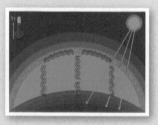
6. _____

2 Discuss the following questions with your partner.

1. What is air pollution? What can possibly cause air pollution?
2. What can we do to reduce air pollution?

Air pollution: causes and solutions

Text A

1. Air pollution is a gas, or a liquid or solid dispersed in a big enough quantity to harm the health of people or other animals, killing plants or stopping them growing properly, damaging or disrupting some other aspects of the environment (such as making buildings crumble), or causing some other kinds of nuisance (reduced visibility, perhaps, or an unpleasant odor).

2. Air pollution is a huge problem — and not just for people living in smog-choked cities: through such things as global warming and damage to the ozone layer, it has the potential to affect us all. So, what exactly causes this major environmental issue, and what can we do about it?

3. Anything people do that involves burning things (combustion), using household or industrial chemicals (substances that cause chemical reactions and may release toxic gases in the process), or producing large amounts of dust has the potential to cause air pollution.

4. By far the biggest culprit today is traffic, though power plants and factories continue to make an important contribution. Now let's look a bit more closely at the three key sources of air pollution.

5. There are over one billion cars on the road today. Many of them are powered by gasoline and diesel engines that burn petroleum to release energy. Petroleum is made up of hydrocarbons (large molecules built from hydrogen and carbon) and, in theory, burning them fully with enough oxygen should produce nothing worse than carbon dioxide and water. In practice, fuels aren't pure

hydrocarbons and engines don't burn them cleanly. As a result, exhausts from engines contain all kinds of pollutants, notably particulates (soot of various sizes), carbon monoxide (CO, a poisonous gas), nitrogen oxides (NOx), volatile organic compounds (VOCs), and lead — and indirectly produce ozone. Mix this noxious cocktail together and energize it with sunlight and you get the sometimes brownish, sometimes blueish fog of pollution we call smog, which can hang over cities for days on end.

6 Renewable energy sources such as solar panels and wind turbines are helping us generate a bigger proportion of our power every year, but the overwhelming majority of electricity is still produced by burning fossil fuels such as coal, gas, and oil, mostly in conventional power plants. Just like car engines, power plants should theoretically produce nothing worse than carbon dioxide and water; in practice, fuels are dirty and they don't burn cleanly, so power plants produce a range of air pollutants, notably sulfur dioxide, nitrogen oxides, and particulates. Power plants also release huge amounts of carbon dioxide, a key cause of global warming and climate change when it rises and accumulates in the atmosphere.

7 Industrial plants and factories that produce the goods we all rely on often release small but significant quantities of pollution into the air. Industrial plants that produce metals such as aluminum and steel, refine petroleum, produce cement, synthesize plastic, or make other chemicals are among those that can produce harmful air pollution. Most plants that pollute release small amounts of pollution continually over a long period of time, though the effects can be cumulative. Sometimes industrial plants release huge of amounts of air pollution accidentally in a very short space of time.

8 Although traffic, power plants, and industrial and chemical plants produce the majority of Earth's outdoor air pollution, many other factors contribute to the problem. In some parts of the world, people still rely on burning wood fuel for their cooking and heating, and that produces indoor air pollution that can

seriously harm their health. In some areas, garbage is incinerated instead of being recycled or landfilled and that can also produce significant air pollution unless the incinerators are properly designed to operate at a high enough temperature.

9. Generally, air pollution is tackled by a mixture of technological solutions, laws and regulations, and changes in people's behavior.

10. **Technological solutions.** Solving air pollution is a challenge because many people have a big investment in the status quo. For example, it's easier for car makers to keep on making gasoline engines than to develop electric cars or ones powered by fuel cells that produce less pollution. The world has thousands of coal-fired power plants and hundreds of nuclear power stations and, again, it's easier to keep those going than to create an entirely new power system based on solar panels, wind turbines, and other forms of renewable energy. Growing awareness of problems such as air pollution and global warming is slowly forcing a shift to cleaner technologies, but the world remains firmly locked in its old, polluting ways.

11. Let's be optimistic, though. Just as technology has caused the problem of air pollution, so it can provide solutions. Cars with conventional gasoline engines are now routinely fitted with catalytic converters that remove some of the pollutants from the exhaust gases. Power plants are fitted with electrostatic smoke precipitators that use static electricity to pull dirt and soot from the gases that drift up smokestacks; in time, it's likely that many older power plants will also be retrofitted with carbon capture systems that trap carbon dioxide to help reduce global warming. Many factories have invested billions of dollars on pollution treatment every year, for instance, dismantling thousands of coal-fired boilers and creating clean heating systems for millions of households. As a replacement for the traditional wood or garbage burning appliances, pellet stoves reduce cost of heating and produce less smoke and ash, since pellets are made from recycled sawdust, which is greatly compressed to burn hotter as

well as cleaner. On a much smaller scale, environmentally friendly people who want to ventilate their homes without opening windows and wasting energy can install heat-recovery ventilation systems, which use the heat energy locked in outgoing waste air to warm fresh incoming air. Technologies like this can help us live smarter — to go about our lives in much the same way with far less impact on the planet.

12 **Laws and regulations.** By itself, technology is as likely to harm the environment as to help it. That's why laws and regulations have been such an important part of tackling the problem of pollution. Many once-polluted cities now have relatively clean air and water, largely thanks to anti-pollution laws.

13 National laws are of little help in tackling transboundary pollution (when air pollution from one country affects neighboring countries or continents), but that doesn't mean the law is useless in such cases. The creation of the European Union has led to many Europe-wide environmental acts, called directives. These force the member countries to introduce their own, broadly similar, national environmental laws that ultimately cover the entire European region.

14 **Raising awareness and changing behavior.** Clean technologies can tackle dirty technologies, and laws can make polluters clean up their act — but none of this would happen without people being aware of pollution and its damaging effects. Sometimes it takes horrific tragedies (like the The Great Smog of 1952 in London or the Chernobyl catastrophe) to prompt action. Often, we pollute the environment without even realizing it: how many people, for example, know that taking a shower or ironing a shirt can release indoor air pollution that they immediately breathe in? Helping people to understand the causes and effects of pollution and what they can do to tackle the issue are very important.

15 Air pollution isn't a single person's problem: it is all of us that help to cause it and we can all help to clean it up.

New words and expressions

disperse /dɪˈspɜːs/ *v.*
to distribute loosely; to cause to separate and go in different directions 分散；传播

combustion /kəmˈbʌstʃən/ *n.*
a process in which substances react with oxygen to give heat and light 燃烧

toxic /ˈtɒksɪk/ *adj.*
of or relating to a toxin or poison 有毒的

culprit /ˈkʌlprɪt/ *n.* 罪魁祸首

exhaust /ɪɡˈzɔːst/ *n.*
the gases coming from an engine as waste products 废气

particulate /pəˈtɪkjʊlət/ *n.*
a substance that consists of very small separate parts 微粒；粒状物

soot /suːt/ *n.*
the black powder which rises in the smoke from a fire 煤烟灰

volatile /ˈvɒlətaɪl/ *adj.* 易挥发的

noxious /ˈnɒkʃəs/ *adj.*
poisonous or harmful 有毒的，有害的

cement /sɪˈment/ *n.* 水泥；结合剂

synthesize /ˈsɪnθɪsaɪz/ *v.*
to produce sth. by combining different things or substances 合成

cumulative /ˈkjuːmjʊlətɪv/ *adj.*
increasing by successive addition 累积的

incinerate /ɪnˈsɪnəreɪt/ *v.*
to burn sth. completely 焚烧

incinerator /ɪnˈsɪnəreɪtə/ *n.* 焚化炉

catalytic /ˌkætəˈlɪtɪk/ *adj.* 催化作用的

electrostatic /ɪˌlektrəˈstætɪk/ *adj.* 静电的

precipitator /prɪˈsɪpɪteɪtə(r)/ *n.* 静电除尘器

retrofit /ˈretrəʊfɪt/ *v.* 翻新

dismantle /dɪsˈmæntl/ *vt.*
to take apart a machine or piece of equipment 拆开，拆卸（机器或设备）

pellet /ˈpelɪt/ *n.*
a small round piece of substance 小团；颗粒状物

sawdust /ˈsɔːdʌst/ *n.*
very small pieces of wood powder 锯末

ventilate /ˈventɪleɪt/ *vt.*
to allow fresh air to enter a place 通风

directive /dɪˈrektɪv; daɪˈrektɪv/ *n.* 指令

Chernobyl catastrophe /tʃɪəˈnɔːbɪl kəˈtæstrəfɪ/ 切尔诺贝利核灾难

diesel engine 柴油机

nitrogen oxide /ˈnaɪtrədʒən ˈɒksaɪd/ 氮氧化物

solar panel 太阳能板

wind turbine /ˈtɜːbaɪn/ 风力涡轮机

sulfur dioxide /ˈsʌlfə daɪˈɒksaɪd/ 二氧化硫

status quo /ˈsteɪtəs kwəʊ/ 现状

Reading comprehension

Read Text A and complete the chart to get the outline and main idea of the text.

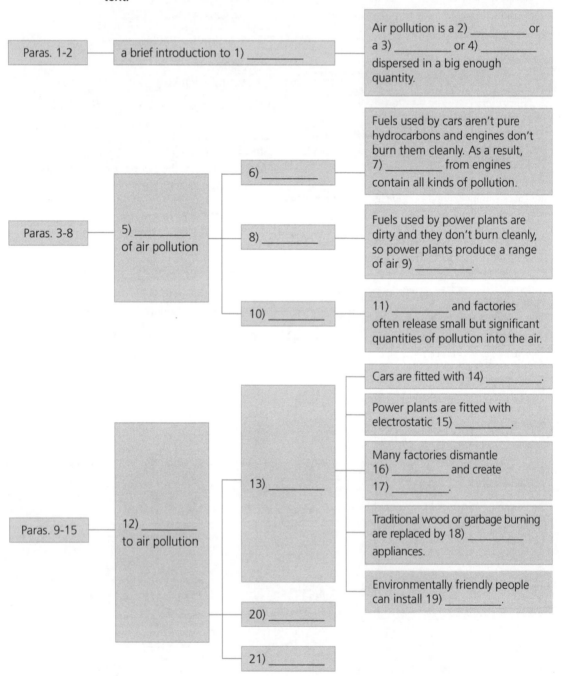

Language focus

1 Match the words and phrases in Column A with the definitions in Column B and give their Chinese meanings in Column C.

Column A	Column B	Column C
___ 1. toxic	a. a colorless gas of which the molecule is composed of oxygen atoms	_____
___ 2. particulate		_____
___ 3. soot	b. containing or relating to poison	_____
	c. to take (sth.) to pieces	
___ 4. odor	d. matter in the form of very small parts	_____
___ 5. ozone	e. a substance that is harmful to the environment	_____
___ 6. pollutant	f. to make sth. by combining different things or substances	_____
___ 7. dismantle		_____
___ 8. synthesize	g. a particular and distinctive smell	_____
	h. black powder which rises in the smoke	

2 Fill in the blanks with the words given below. Change the form where necessary.

combustion   convert   disperse   exhaust
incinerate   pollute   precipitate   ventilate

1. Diesel _____ is believed to consist of thousands of organic constituents and is a major cause of urban pollution.
2. By cleaning up the pollutants left over from burning, catalytic _____ reduce vehicles of tailpipe emissions of hydrocarbons and carbon monoxide to extremely low levels.
3. Air pollution refers to the release of _____ into the air that are detrimental to human health.
4. In less enlightened times, factory operators thought that if they built really high smokestacks, the wind would simply blow their smoke away, diluting and _____ it so it wouldn't be a problem.

5. Indoor air pollution can be reduced by making sure that a building is _____ and cleaned regularly to prevent the buildup of agents like dust and mold.
6. The 20th century saw the popular acceptance of the automobile and the internal _____ engine, which led to the pollution of the air.
7. Most solid waste is buried in sanitary landfills, but small percentage of municipalities _____ their refuse.
8. Known as scrubbers, the electrostatic smoke _____ snatch the soot and ash from dirty air as they flow along a pipe, greatly reducing pollution and helping to improve the environment.

3 Read the text and decide whether the following statements are true or false. If false, underline the mistakes and put the corrections in the blanks provided.

1. Air pollution is a gas released in a big enough quantity to cause a lot of damage to the health of aminals, plants, and the environment. _____
2. Many cars today are powered by gasoline and diesel engines that burn petroleum to release energy. _____
3. Burning petroleum fully with enough oxygen should produce nothing worse than carbon monoxide and water. _____
4. Exhausts from engines contain all kinds of pollutants, notably particulates (soot of various sizes), carbon monoxide (CO, a poisonous gas), nitrogen oxides (NOx), volatile organic compounds (VOCs), and lead — and directly produce ozone. _____
5. Mix the engine exhausts together and energize it with sunlight, you will get the fog of pollution named ozone, which can hang over cities for days on end. _____
6. Most plants that pollute release small amounts of pollution continually over a long period of time, though the effects can be digressive. _____
7. The world has thousands of coal-fired power plants and hundreds of nuclear power stations and it's easier to keep those going than to create an entirely new power system based on solar panels, wind turbines, and other forms of renewable energy. _____

8. Power plants are fitted with electrostatic smoke precipitators that use static electricity to pull NOx from the gases that drift up smokestacks.
   _____

9. Many factories will be retrofitted with carbon capture systems that trap carbon dioxide to help reduce global warming.
   _____

10. Environmentally friendly people who want to ventilate their homes without opening windows and wasting energy can install heat-recovery ventilation systems. _____

4 Translate the following paragraph into English.

地球大气层由一层气体组成，这些气体能吸收足够的热量使生命得以维持。然而，因为燃烧化石燃料和砍伐森林，人类向大气中排放了数十亿吨的二氧化碳。此外，我们还向大气中排放了少量的其他气体，这些气体吸收了更多来自太阳光线的热量，加剧了温室效应。因此，全球气温不断上升，导致全世界的气候失衡。虽然许多温室气体是自然产生的，但人类向大气中排放温室气体的速度远超自然。据估计，目前二氧化碳的浓度比工业革命之前高出三分之一以上，而工业革命正是大规模燃烧化石燃料和现代工农业的开端。必须采取各种行之有效的措施来遏制全球大气变暖。

Critical thinking

1 Some people think that the environment can be self-cleaning when pollution reaches a certain level of complexity, so we can depend on the environment to clear up air pollutants. What is your understanding? Work in groups and discuss your thoughts.

2 According to Text A, the problem of air pollution should be tackled by a comprehensive solution of three aspects: technological solutions, laws and regulations, and changes in people's behavior. What would happen if one of these aspects is taken out?

Research task

Academic skill: collecting data

Data collection is one of the most important stages in conducting research. Accurate and systematic data collection is critical to scientific research. There are many methods to collect data, depending on the design of research and the methodologies employed. Some of the common methods are questionnaire, interview and observation.

1. How to design a questionnaire

    A questionnaire can be designed for descriptive and analytical surveys. In a descriptive survey, the questionnaire will normally use nominal and ordinal scales because it concerns primarily with the particular characteristics of a specific subject.

    What is your gender?
    ☐ Male   ☐ Female

    What is your hair color?
    ☐ Brown   ☐ Black   ☐ Blonde   ☐ Gray   ☐ Other

    Example of nominal scale:

    How do you feel today?
    ☐ Very unhappy   ☐ Unhappy   ☐ OK
    ☐ Happy   ☐ Very happy

    How satisfied are you with our service?
    ☐ Very unsatisfied   ☐ Somewhat unsatisfied   ☐ Neutral
    ☐ Somewhat satisfied   ☐ Very satisfied

    Example of ordinal scale:

    Rating scale is always applied to measure the attitude or opinion of the respondents in an analytical survey. The most popular one is Likert scale. Usually you would use a 1-5 rating scale where: 1 = strongly disagree; 2 = somewhat disagree; 3 = undecided; 4 = somewhat agree; 5 = strongly agree.

Example of Likert scale — the employment self-esteem scale:

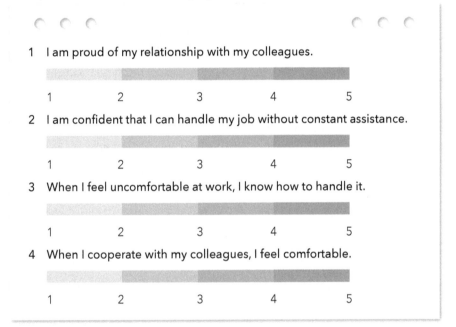

When designing a questionnaire, you have to pay attention to the following issues:
- Are the instructions clear and unambiguous?
- Can the questions be understood? Are they free from jargon, terminology, unsuitable assumption and ambiguity?
- Do the respondents have the necessary knowledge to answer the questions?
- Do the questions appear offensive or embarrassing to the respondents?
- Do the questions lead the respondents to particular answers?

2. How to conduct an interview

Since an interview involves two-way communication, there are certain rules and guidelines to be followed:
- Ask one question at a time.
- Remain as unbiased as possible. Don't show strong emotional reactions to the responses of the interviewee.

- Verify understanding through raising and confirming questions.
- Let the interviewee do most of the talking.
- Maintain control over the subject matter.

3. How to conduct observation

There are generally two ways of conducting observation, namely non-participant observation and participant observation. The researcher in non-participant observation does not involve in the studied subjects. Data are collected by observing the behavior or phenomenon. In contrast, the researcher in participant observation immerses into ongoing activities and makes observation records. Data are collected by interacting with the studied subjects. Here are some tips for conducting observation:

- The collection of detailed field notes is key to successful observation.
- Audio recorders or cameras can be used to aid with capturing raw data.
- Researchers in participant observation should state their intentions openly.
- Researchers in non-participant observation should adopt a more separate and distant role than that of the participant observers.
- Non-participant observation can be overt or covert.

Task

Please design a questionnaire about your schoolmates' understanding of global warming and their attitudes toward it.

The questions can be designed using

1) nominal scales, for example, *please state your major*;
2) ordinal scales, for example, *do you think that fossil fuels will be replaced by renewable resources such as solar energy: very optimistic, somewhat optimistic, neutral, somewhat pessimistic, very pessimistic*;
3) rating scales, for example,

	Strongly disagree				Strongly agree
It is high time to stop dependence on fossil fuels.	1	2	3	4	5

Section B

Reading strategy

Dealing with unknown words (Part II)

Guessing by context

This is a more advanced skill. This time you think about the general meaning of the sentence and make a guess at the probable meaning of that word. While you are only guessing, it is a skill that does improve with practice: the more you guess meanings, the more correct you are. For example:

> "By far the biggest culprit today is traffic, though power plants and factories continue to make an important contribution. Now let's look a bit more closely at the three key sources of air pollution."

The word "culprit" may be a word that you are not familiar with. However, if you read the next sentence, you can easily get its meaning: something responsible for causing a problem.

Guessing by affix

Analyzing the word formation can help you determine the meaning of an unfamiliar word. What you should do here is recognize the parts of words and relate them to other words you know. Again, this will mean you are "guessing" and sometimes you may make mistakes, but you should be correct more often than not. For example:

> "National laws are of little help in tackling transboundary pollution (when air pollution from one country affects neighboring countries or continents), but that doesn't mean the law is useless in such cases."

"Transboundary" may be a new word to you, but you can guess its meaning from "trans-" and "boundary". The prefix "trans-" means "across", "beyond", or "on the opposite side", and "boundary" means "a dividing line" or "border". Combined with the context, you can see it must mean: across national borders.

Guessing by the part of speech of a word

Guessing by the part of speech of a word can sometimes help to know whether you are looking at a verb, noun, adverb or adjective. For example:

> "On a much smaller scale, environmentally friendly people who want to ventilate their homes without opening windows and wasting energy can install heat-recovery ventilation systems, which use the heat energy locked in outgoing waste air to warm fresh incoming air."

"Ventilate" may be strange to you. From the text, you can tell it must be a verb as it follows the infinitive "to". And, with the help of other clues like "opening windows" and "ventilation systems which use the heat energy locked in outgoing waste air", you might guess by logical reasoning that "ventilate" might mean "to let air in and out".

Task

Now read Text B and apply the skills above to deal with the underlined words. Remember: do not turn to a dictionary directly when coming across an unfamiliar word, but try to "guess" the meaning of it.

How air pollution works on different scales

1. Air pollution can happen on every scale, from local to global. Sometimes the effects are immediate and happen very near to the thing that caused them; but they can also appear days, months, or even years later — and in other cities, countries, or continents.

Local air pollution

2. Have you ever sat on a train with someone who suddenly decided to start cleaning or varnishing their nails? Acetone (a solvent in nail varnish remover) is a VOC (volatile organic compound), so it evaporates and spreads very quickly, rapidly getting up the nose of anyone sitting nearby. Open a can of gloss paint in your home and start painting a door or window and your house will very quickly be filled with a noxious chemical stench – VOCs again! Grill

---

solvent *n.* 除臭剂
varnish *n.* 指甲油

some toast too long and you'll set the bread on fire, filling your kitchen with clouds of soot (particulates) and possibly setting off a smoke alarm or carbon monoxide detector. These are three everyday examples of how air pollution can work on a very local scale: the causes and the effects are close together in both space and time. Localized air pollution like this is the easiest kind to tackle.

3 A case of local air pollution is the indoor air pollution. If you live in a city, you might think your home is the cleanest place you can be — but you're probably wrong. Outside, though the air may seem polluted, it's constantly moving and (in theory at least) pollutants are continually being diluted and dispersed. Inside, your home is packed with all kinds of chemicals that generate pollution every time you use them. And, unless you open the windows regularly, those pollutants aren't going anywhere fast. There are many kinds of home pollutions, some of which you might have never noticed before:

- Detergents and household cleaners, aerosol sprays, shoe polish, hair wax, paints, and glues are just a few of the everyday chemicals that can release air pollution into your home.
- If you have a gas or oil-fired boiler or a coal- or wood-fired stove and it's not properly ventilated, it will generate dangerous and toxic (but colorless and odorless) carbon monoxide gas.
- Surprisingly, even the water that pipes into our homes can be a source of air pollution. Every time you heat water (on a stove, in a kettle, in a shower, or even when you're steam ironing clothes), you can evaporate VOC chemicals trapped inside and release them into the air.
- Even your shiny new shower curtain could be releasing VOCs if it's made from a type of plastic called PVC.
- Maybe your building has air conditioning? Chances are the air it blows through has already circulated through other rooms in the same building or

---

diluted *adj.* 被稀释的
detergent *n.* 清洁剂；洗涤剂
aerosol spray 气雾喷雾剂

even other people's offices or apartments.
- Perhaps your building is located somewhere near a source of natural radioactivity so radon gas is slowly accumulating inside.

4 None of these things are meant to scare you — and nor should they. Just remember that there's pollution inside your home as well as outside and keep the building well ventilated. (If you're worried about wasting energy by opening windows on cold days, there are systems that can let air into a building without letting the heat escape, known as heat-recovery ventilation.)

Neighborhood air pollution

5 How clean your air is depends on where you live: air is generally far cleaner in rural than in urban areas, for example, where factories, chemical plants, and power plants are more likely to be located and traffic levels are much higher. Exactly how clean your neighborhood is can also depend critically on the weather, especially if you live somewhere prone to temperature inversions and smog. Neighborhood air pollution problems are often best tackled through local community campaigns.

Regional air pollution

6 Tall smokestacks designed to disperse pollution don't always have that effect. If the wind generally blows in the same direction, the pollution can be systematically deposited on another city, region, or country downwind.

7 It's often said that pollution knows no boundaries — and that's particularly true of air pollution, which can easily blow from one country or continent where it's produced and cause a problem for someone else. Air pollution that travels like this, from country to country, is called transboundary pollution; acid rain is also an example of this and so is radioactive fallout (the contaminated dust that falls to Earth after a nuclear explosion). When the Chernobyl nuclear power plant

---

radon *n.* 氡

exploded in the Ukraine in 1986, wind dispersed the air pollution it produced relatively quickly — but only by blowing a cloud of toxic radioactive gas over much of Europe and causing long-lasting problems in a number of other countries (70 percent of the fallout landed on neighboring Belarus).

Global air pollution

8 It's hard to imagine doing anything so dramatic and serious that it would damage our entire, enormous planet — but, remarkable though it may seem, we all do things like this every day, contributing to problems such as global warming and the damage to the ozone layer (two separate issues that are often confused).

9 Every time you ride in a car, turn on the lights, switch on your TV, take a shower, microwave a meal, or use energy that's from burning a fossil fuel such as oil, coal, or natural gas, you're almost certainly adding to the problem of global warming and climate change: unless it's been produced in some environmentally friendly way, the energy you're using has most likely released carbon dioxide gas into the air. While it's not an obvious pollutant, carbon dioxide has gradually built up in the atmosphere, along with other chemicals known as greenhouse gases. Together, these gases act a bit like a blanket surrounding our planet that is slowly making the mean global temperature rise, causing the climate (the long-term pattern of our weather) to change, and producing a variety of different effects on the natural world, including rising sea levels.

10 Global warming is a really dramatic effect of air pollution produced by humans, but that doesn't mean it's an insoluble problem. People have already managed to solve another huge air pollution problem that affected the whole world: the damage to a part of the atmosphere called the ozone layer. At ground level, ozone is an air pollutant — but the ozone that exists in the stratosphere (high up in the atmosphere), is exactly the opposite: it's a perfectly natural

Belarus 白俄罗斯

chemical that protects us like sunscreen, blocking out some of the Sun's harmful ultraviolet radiation. During the 20th century, people started using large quantities of chemicals called chlorofluorocarbons (CFCs), because they worked very well as cooling chemicals in refrigerators and propellant gases in aerosol cans (propellants are the gases that help to fire out air freshener, hair spray, or whatever else the can contains). In 1974, scientists Mario Molina and Sherwood Rowland suggested that chlorofluorocarbons attacked and destroyed the ozone layer, producing holes that would allow dangerous ultraviolet light to stream through. In the 1980s, huge "ozone holes" started to appear over Antarctica, prompting many countries to unite and sign an international agreement called the Montreal Protocol, which rapidly phased out the use of CFCs. As a result, the ozone layer — though still damaged — is expected to recover by the end of the 21st century.

Unit 3

Water pollution

In this unit, you will learn:

- **Subject-related knowledge:** The definition and classification of water pollution
  Solutions to water pollution
- **Academic skill:** Writing a literature review
- **Reading strategy:** Identifying the main information in a sentence

Section A

Pre-reading

1 Match the expressions with the pictures below.

municipal waste water   livestock farm waste water   agricultural waste water
industrial waste water   power plant hot water   mine waste water

1. _____   2. _____   3. _____

4. _____   5. _____   6. _____

2 Discuss the following questions in groups.

1. What do you know about water pollution? What causes water pollution?
2. What can we do to reduce water pollution around us?

Water pollution: an introduction

1. Over two thirds of Earth's surface is covered by water; less than a third is taken up by land. As Earth's population continues to grow, people are putting ever-increasing pressure on the planet's water resources. In a sense, our oceans, rivers, and other inland waters are being "squeezed" by human activities – not so they take up less room, but so their quality is reduced. Poorer water quality means water pollution.

2. Water pollution can be defined in many ways. Usually, it means one or more substances have built up in water to such an extent that they cause problems for animals or people. Oceans, lakes, rivers, and other inland waters can naturally clean up a certain amount of pollution by dispersing it harmlessly. If you poured a cup of black ink into a river, the ink would quickly disappear into the river's much larger volume of clean water. The ink would still be there in the river, but in such a low concentration that you would not be able to see it.

At such low levels, the chemicals in the ink probably would not present any real problem. However, if you poured gallons of ink into a river every few seconds through a pipe, the river would quickly turn black. The chemicals in the ink could very quickly have an effect on the quality of the water. This, in turn, could affect the health of all the plants, animals, and humans whose lives depend on the river.

3 Thus, water pollution is all about quantities: how much of a polluting substance is released and how big a volume of water it is released into. A small quantity of a toxic chemical may have little impact if it is spilled into the ocean from a ship. But the same amount of the same chemical can have a much bigger impact pumped into a lake or river, where there is less clean water to disperse it.

4 Water pollution almost always means that some damage has been done to an ocean, river, lake, or other water sources. A 1969 United Nations report defined ocean pollution as: "The introduction by man, directly or indirectly, of substances or energy into the marine environment (including estuaries) resulting in such deleterious effects as harm to living resources, hazards to human health, hindrance to marine activities, including fishing, impairment of quality for use of sea water and reduction of amenities."

5 Fortunately, Earth is forgiving and damage from water pollution is often reversible.

6 When we think of Earth's water resources, we think of huge oceans, lakes, and rivers. Water resources like these are called surface waters. The most obvious type of water pollution affects surface waters. For example, a spill from an oil tanker creates an oil slick that can affect a vast area of the ocean.

7 Not all of Earth's water sits on its surface, however. A great deal of water is held in underground rock structures known as aquifers, which we cannot see and seldom think about. Water stored underground in aquifers is known as groundwater. Aquifers feed our rivers and supply much of our drinking water.

They too can become polluted, for example, when weed killers used in people's gardens drain into the ground. Groundwater pollution is much less obvious than surface-water pollution, but is no less of a problem. In 1996, a study in Iowa in the United States found that over half the state's groundwater wells were contaminated with weed killers.

8 Surface waters and groundwater are the two types of water resources that pollution affects. There are also two different ways in which pollution can occur. If pollution comes from a single location, such as a discharge pipe attached to a factory, it is known as point-source pollution. Other examples of point-source pollution include an oil spill from a tanker, a discharge from a smokestack (factory chimney), or someone pouring oil from their car down a drain. A great deal of water pollution happens not from one single source but from many different scattered sources. This is called non-point-source pollution.

9 When point-source pollution enters the environment, the place most affected is usually the area immediately around the source. For example, when a tanker accident occurs, the oil slick is concentrated around the tanker itself and, in the right ocean conditions, the pollution disperses further away from the tanker. This is less likely to happen with non-point-source pollution which, by definition, enters the environment from many different places at once.

10 Sometimes pollution that enters the environment in one place has an effect hundreds or even thousands of miles away. This is known as trans-boundary pollution. One example is the way radioactive waste travels through the oceans from nuclear reprocessing plants in England and France to nearby countries such as Ireland and Norway.

11 Some people believe pollution is an inescapable result of human activity: They argue that if we want to have factories, cities, ships, cars, oil, and coastal resorts, some degree of pollution is almost certain to result. In other words, pollution is a necessary evil that people must put up with if they want to make

progress. Fortunately, not everyone agrees with this view. One reason people have woken up to the problem of pollution is that it brings costs of its own that undermine any economic benefits that come about by polluting.

12 Take oil spills for example. They can happen if tankers are too poorly built to survive accidents at sea. But the economic benefit of compromising on tanker quality brings an economic cost when an oil spill occurs. The oil can wash up on nearby beaches, devastate the ecosystem, and severely affect tourism. The main problem is that the people who bear the cost of the spill (typically a small coastal community) are not the people who caused the problem in the first place (the people who operate the tanker). Yet, arguably, everyone who puts gasoline (petrol) into their car – or uses almost any kind of petroleum-fueled transport – contributes to the problem in some way. So oil spills are a problem for everyone, not just people who live by the coast and tanker operators.

13 Pollution matters because it harms the environment on which people depend. The environment is not something distant and separate from our lives. It's not a pretty shoreline hundreds of miles from our homes or a wilderness landscape that we see only on TV. The environment is everything around us that gives us life and health. Destroying the environment ultimately reduces the quality of our own lives — and that, most selfishly, is why pollution should matter to all of us.

14 Life is ultimately about choices — and so is pollution. We can live with sewage-strewn beaches, dead rivers, and fish that are too poisonous to eat. Or we can work together to keep the environment clean so the plants, animals, and people who depend on it remain healthy. We can take individual action to help reduce water pollution, for example, by using environmentally friendly detergents, not pouring oil down drains, reducing pesticides, and so on. We can take community action too, by helping out on beach cleaning or litter picking to keep our rivers and seas a little bit cleaner. And we can take action as countries and continents to pass laws that will make pollution harder and the world less polluted. Working together, we can make pollution less of a problem – and the world a better place.

New words and expressions

squeeze /skwiːz/ *vt.*
to press sth., especially with your fingers 捏，挤

substance /ˈsʌbstəns/ *n.*
a type of solid, liquid or gas that has particular qualities 物质

spill /spɪl/
vt. to accidentally pour a liquid out of its container（意外地）（使）溢出，（使）泼出，（使）洒出
n. an act of spilling or an amount of sth. that is spilled 溢出；溢出量

pump /pʌmp/ *vt.*
to make liquid or gas move into or out of sth., especially by using a pump（尤指用泵）注入，抽取

estuary /ˈestjuəri/ *n.* 河口湾；港湾

deleterious /ˌdelɪˈtɪəriəs/ *adj.*
harmful and damaging 有害的；造成伤害的

hindrance /ˈhɪndrəns/ *n.*
the act of making it difficult for sb. to do sth. 妨碍；阻挠

impairment /ɪmˈpeə(r)mənt/ *n.*
the condition of being damaged, or being weaker or worse than usual 受损；变差

amenity /əˈmiːnəti/ *n.* 便利设施；娱乐设施

reversible /rɪˈvɜːrsəbl/ *adj.*
able to return or be changed to a previous state 可逆转的；可倒转的

tanker /ˈtæŋkə(r)/ *n.* 油罐车；油船，油轮

aquifer /ˈækwɪfə(r)/ *n.*
a layer of stone or earth under the surface of the ground that contains water 地下蓄水层

drain /dreɪn/
vi. if liquid drains, it flows away （液体）流走，流失
n. a pipe or hole that dirty water or waste liquids flow into 下水管；下水道；排（废）水管

resort /rɪˈzɔːt/ *n.*
a place that is popular for recreation and vacations and provides accommodations and entertainment 度假胜地

undermine /ˌʌndəˈmaɪn/
vt. to make sth. or sb. become gradually less effective, confident, or successful 逐步削弱；逐渐损害

ecosystem /ˈiːkəʊˌsɪstəm/ *n.*
all the plants and living creatures in a particular area considered in relation to their physical environment 生态系统

gasoline /ˈɡæsəliːn/ *n.* 汽油

sewage /ˈsjuːɪdʒ/ *n.*
used water and waste substances that are produced by human bodies, that are carried away from houses and factories through special pipes（下水道的）污水；污物

detergent /dɪˈtɜːdʒənt/ *n.*
a liquid or powder that helps remove dirt, for example from clothes or dishes 洗涤剂；去垢剂；洗衣粉

pesticide /ˈpestɪsaɪd/ *n.*
a chemical used for killing pests, especially insects 杀虫剂；除害药物

build up to increase or develop gradually（使）增加；（使）增强

clean up to remove pollution from a place or an industrial process 清除（污染）

oil slick a layer of oil floating on the ocean or on a lake 浮油

point-source pollution 点源污染

wash up
if the sea washes sth. up somewhere, it carries it and leaves it there（海浪）把（某物）冲上岸；（某物）被海浪冲上岸

petroleum-fueled *adj.* 石油燃料的

sewage-strewn *adj.* 污水扩散的

Reading comprehension

1 Text A can be divided into five parts. Now write down the paragraph number(s) of each part and then give the main idea in one or two sentences.

Part	Paragraph(s)	Main idea
I	Para(s). _____	
II	Para(s). _____	
III	Para(s). _____	
IV	Para(s). _____	
V	Para(s). _____	

Language focus

1 Match the words in bold in Column A with their corresponding definitions or explanations in Column B.

Column A	Column B
___ 1. Oceans, lakes, rivers, and other inland waters can naturally clean up a certain amount of pollution by **dispersing** it harmlessly.	A. poisonous, or containing poison
___ 2. The ink would still be there in the river, but in such a low **concentration** that you would not be able to see it.	B. the amount of a substance in a liquid or in another substance
___ 3. But the economic benefit of compromising on tanker quality brings an economic cost when an oil **spill** occurs.	C. a chemical substance in the form of a powder or a liquid for removing dirt from clothes or dishes, etc.
___ 4. In 1996, a study in Iowa in the United States found that over half the state's groundwater wells were **contaminated** with weed killers.	D. to make sth. dirty, polluted, or poisonous by adding a chemical, waste, or infection
___ 5. A small quantity of a **toxic** chemical may have little impact if it is spilled into the ocean from a ship.	E. the amount of liquid that comes or falls out of a container
___ 6. We can take individual action to help reduce water pollution, for example, by using environmentally friendly **detergents**, not pouring oil down drains, reducing pesticides, and so on.	F. to cause sth. to scatter in different directions, or scatter in this way

2 Complete the following sentences with the expressions given below. Change the form where necessary.

take action   take up   pump into   wash up
live with   drain into spill into   clean up

1. Agricultural irrigation _____ a large amount of freshwater resources in some areas with severe water shortage.
2. Thousands of volunteers and experts are trying to devise ways to _____ the huge slick.
3. Now, waste water from these farms is often just _____ nearby pits.
4. The removal of forest cover may also have allowed rainwater to _____ the rivers.
5. The country has been troubled by the threat of a huge slick _____ on its beaches.
6. We hope this information will make them actively _____ to protect the water resources.
7. When _____ the sea, oil can be toxic to marine plants and animals.
8. In many developing countries, people have to _____ sewage-strewn beaches, dead rivers, and fish that are too poisonous to eat.

3 Translate the following terms into Chinese.

1. a polluting substance   _____
2. a small quantity of a toxic chemical   _____
3. the marine environment   _____
4. deleterious effects   _____
5. living resources   _____
6. an oil slick   _____
7. surface-water pollution   _____
8. a discharge pipe attached to a factory   _____
9. point-source pollution   _____
10. a discharge from a smokestack   _____
11. radioactive waste   _____
12. nuclear reprocessing plants   _____
13. petroleum-fueled transport   _____

14. sewage-strewn beaches　　　　_____
15. environmentally friendly detergents　　_____

4 Translate the following paragraph into English.

近年来，这条河的污染问题与洪水问题一样令人担忧。它的水体40%都被严重污染，甚至不能用于灌溉。之后对水质进行了改善，但是到2017年，从河里提取的水样中仍有十分之一不适合农用。从那以后，引入了中央数字控制系统来控制大坝放水。这一被列为世界上最先进的水资源配给系统保障了河水不间断地流入海洋。

Critical thinking

1 The following pie chart shows the effects of water pollution on human life. Discuss in groups the interrelationship between water pollution and human life.

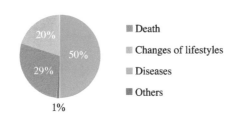

The effects of water pollution on human life

- Death
- Changes of lifestyles
- Diseases
- Others

2 What do you think should the people facing the problem of water pollution do to improve their situation? Please write a draft for your speech and present it to the class.

Research task

Academic skill: writing a literature review

What is a literature review?

A literature review is an account of what has been published on a topic by accredited scholars or researchers. The purpose of writing it is to convey to the reader what knowledge or ideas have been established on that topic, and what strengths and weaknesses those studies have.

A literature review is usually intended to summarize and synthesize the previous works on the topic. A summary is a recap of the important information on the topic, while a synthesis is a re-organization, or a reshuffling of the information. A literature review might give a new interpretation of old materials or combine new interpretations with old ones. It might also trace the intellectual progression of the field, including major debates.

How do you write a literature review?

First, focus on your specific topic. If your topic is too broad, you will have far too many sources to review. Narrow down your task as much as possible while still satisfying the research needs.

Second, find relevant literature. Choose material that aids the understanding and validation of the topic being discussed. Make sure the author is qualified and the argument is convincing.

Third, write a clear, short introduction. Include an overview of the topic or theory being studied, and the objective of the literature review.

Fourth, summarize your studies. In the summary, try to compare and contrast different authors' views on the topic and group the authors who draw similar conclusions.

Finally, synthesize your summary. Synthesize the results based on the summary and indicate what is and is not known, identify areas of controversy in the literature, formulate questions that need further research, and show how your study relates to the previous studies.

Sample

Because of the energy crisis and environmental pollution, energy saving and emission reduction are of vital importance for social development. Large-scale utilization and centralized integration of renewable energy in China become an inevitable trend. However, wind energy and solar energy, as new kinds of renewable energy, are much different from conventional resources, and their inherent characteristics of stochastic volatility and intermittency bring massive challenges to the dispatch and operation of power systems. In the experimental research field, many scholars have drawn attention to large-scale energy storage technologies. At present, some achievements and application practices have been realized in renewable energy generation monitoring such as analyzing energy storage and its use with intermittent renewable energy (Barton JP, Infield DG, 2004), studying a stochastic simulation of battery sizing for demand shifting and uninterruptible power supply facility (Tan CW, Green TC, Hernandez-Aramburo CA, 2007), studying the reliability considerations in the utilization of wind energy, solar energy and energy storage in electric power systems (Billinton RB, 2006), explaining design and realization of computer integrated monitoring system of wind farm (Huang SZ, Dai ZQ, Ma XP, 2010), presenting the study on integrated monitoring & control platform of wind farms (Qiao Y, Lu ZX, Xu F et al, 2011), discussing the design and implementation of photovoltaic power generation monitor control system (Wang CF, Li R, Liu HR et al, 2011), explaining the small-scale wind-solar hybrid (混合物) grid-connected generation system (Zhao M, Lu JZ, Zhou RJ et al, 2011), introducing hybrid energy storage system and control system design for wind power balancing (Yu P, Zhou W, Sun H et al, 2011), and proposing energy management strategy of wind-PV-ES hybrid power system (Li JX, Zhang JC, Zhou Y, 2011). Although these studies have analyzed the storage and monitoring about wind and solar energy, a complete set of implementation scheme with grid-friendly interaction is not put forward. Engineering applications and operating experiences on application of wind, PV and storage co-generation monitoring system are still too few. Further research in models and strategies of multiple-source complementary coordination control is needed.

Task

"Surface-water pollution", mentioned in Text A, is one type of pollution. Work in groups of four. Each of you conducts research on one of the following aspects of surface-water pollution and writes a literature review. Then share your findings in your group and discuss if there are any questions that need further research about this topic.

- causes of surface-water pollution;
- effects of surface-water pollution;
- solutions to surface-water pollution;
- efficiency of these solutions.

Section B

Reading strategy

Identifying the main information in a sentence

Sentences in academic and technical texts are often very long. For example:

But the economic benefit of compromising on tanker quality brings an economic cost when an oil spill occurs.

1) In order to grasp the main idea of the sentence without confusion, first you should identify the main structure — the subject, the verb and the object, if there is one — instead of understanding every word. For example, in the sentence above, you can find:

   subject = economic benefit

   verb = brings

   object = economic cost

   Bear in mind that to find the main structure, you should remove modifiers, e.g. "of compromising on tanker quality" and "when an oil spill occurs".

2) Now you have known the main structure of the sentence, you can get the main idea of it. Long supplementary information is often seen in sentences with complicated structures. Since it takes up a large part of the whole sentence, you might want to know briefly about its function and meaning. In the complement, you should omit the functional words, parenthesis and attributive, which will save you a lot of time trying to read through the details of unfamiliar names or unimportant messages.

Task

Read Text B and apply the skills above to find the main information of the underlined sentences.

Text B

Solutions to water pollution

1. In many countries, there are strong laws to protect clean water. Unfortunately, good starts are not always accompanied by strong follow-through. For instance, implementation of the U.S. Clean Water Act in the 1970s did improve water quality over the subsequent decade or two. However, though the Act called for "zero discharge of pollutants into navigable waters by 1985, and fishable and swimmable waters by 1983", almost all surface waters in the U.S. still suffer some level of pollution, discharges are still permitted, and the EPA (Environmental Protection Agency) still categorizes roughly 40% of lakes, rivers, and streams as unsafe for fishing or swimming.

2. The first water pollution solution is simple: enforce existing laws. A politician pontificating about a great new anti-pollution law they've sponsored means little if they continue to allow existing laws to go unenforced.

3. Beyond laws, there are some practical water pollution solutions that can be implemented by society and by individuals.

Reducing nutrient and pesticide pollution

4. Solutions to water pollution caused by excess nutrients and chemical pesticides

---

pontificate *vi.* 自负地谈论

can be found in four broad categories:

• **Encourage smart agricultural practices**

5 Right-sizing applications of fertilizer and using techniques like biodynamic farming, no-till planting, settling ponds, and riparian buffer zones can help keep polluted runoff from entering streams.

6 Animal agriculture can be a problem too. For example, runoff from chicken-raising operations is a leading cause of nutrient pollution in the Chesapeake Bay. As a compassionate society, we should probably ban altogether the barbaric practices imposed on animals in concentrated feeding operations; but until we do that, these potent pollution centers must be made to meet strict water quality standards.

• **Reduce urban / suburban runoff of lawn fertilizers and pesticides**

7 If you put "normal" fertilizer, pesticides, and other chemicals on your lawn, landscaping, and gardens, you are part of the water pollution problem. While you may find these products helpful, much of their volume is washed by rain or blown by air to nearby streams, ponds, and rivers. They also tend to degrade your soil over time, making your future gardening efforts much more difficult and reliant on chemicals. To tackle the problems, you can use organic, non-chemical fertilizers and try to use natural ways to deal with pests. For example, grow certain plants whose smell they don't like.

• **Prevent further destruction of wetlands, and reestablish them wherever possible**

8 Both inland and coastal wetlands act to buffer surges in runoff and to filter pollutants from runoff and flows. Yet it has been standard practice in the U.S. (and many other countries) to allow development concerns to almost always trump the value of "nature's services". It's time to get serious about preserving wetlands.

no-till planting 免耕栽培
settling pond 沉淀池
runoff *n.* （生产过程中）排出的废水或废物
surge *n.* 激增

- **Improve sewage treatment**

9 If the toilet in your house were spewing its contents onto your bathroom floor, you would make it a very high priority to get the situation corrected. As societies, we should place the same priority on upgrading out-of-date or under-capacity sewage treatment plants that sometimes spew their contents into our waterways. It's also important to ensure that homeowners with septic fields are installing and maintaining their systems in a way that does not contaminate nearby groundwater or surface water.

Stopping deforestation

10 A healthy forest acts like a sponge to soak up the rains when they come, holding the water and filtering it before it makes its way to nearby streams, lakes, and rivers. When all the trees are cut down – clear cutting is still logging companies' preferred method of operation – the forest ecosystem dies and can no longer perform this service. Rain water rushes directly into streams, flowing over exposed soil, picking up and carrying sediment pollution into nearby waterways.

Opposing coastal development

11 Natural shorelines (and the wetlands usually found there) serve many purposes, from fish nurseries to absorption of hurricane impact to filtration of the river water entering the estuary. But in the U.S. alone, more than 20,000 acres of these sensitive areas disappear each year. When houses, hotels, and resorts go up or other development occurs, not only are the wetland and coastal eco-services lost, but the human activity imparts many types of pollution to these sensitive coastal areas.

12 Coastal development is a significant problem for the oceans, but all forms of suburban sprawl chew up wetlands, forests, meadows, and other natural areas

---

spew *vt.* 涌出；渗出
septic *n.* 化粪池
sediment *n.* 沉积物
sprawl *v.* 蔓延；（建筑物等）无计划地延伸

that help soak up rains and filter water before it enters streams and rivers. Supporting smart growth, urban redevelopment, and open space preservation is an important solution to water pollution.

Reducing pollution from oil and petroleum liquids

13   While it's true that a large amount of oil naturally seeps into the ocean from underground geological sources, marine life in the areas where this occurs has had eons to adapt to the conditions. Human-caused petroleum pollution invariably happens in much more sensitive areas, often with disastrous consequences.

14   The first-level solution to this type of water pollution is to stop letting so much oil and oil byproducts get into the water in the first place. Yes, we must reduce the occurrences of oil spills; but more importantly, we must reduce the amount of petroleum pollution getting into waterways from non-spill sources, which contribute far more to the problem than spills.

Reducing mercury emissions

15   It's a real shame that we have let global mercury pollution get so bad that wonderful fish like albacore tuna and swordfish are polluted to the point where they aren't safe to eat, at least not in any significant quantities.

16   The solution to mercury pollution in our waters is to solve the mercury pollution problem coming from the land. In the United States and many other countries, coal-burning power plants are the largest human-caused source of mercury emissions. The U.S. power plants account for over 40% of its mercury emissions. Other noteworthy sources of mercury pollution are chlorine production facilities and municipal and hazardous waste incinerators.

eon *n.* 10^9 年；极长的时间
chlorine *n.* 氯

Reducing chemical pollution

17 Biologist Joseph Sheldon has described the chemical inundation of our biosphere as "global toxification". Chemicals are everywhere, in everything – "better living through chemistry" has turned out to have a serious pollution downside. Here's what we can do about it.

18 First, there are some things we should reinvigorate progress on cleaning up polluted sites, and reestablish the "polluter pays" principle. Taxpayers should not foot the bill for decades of industry abuses.

19 Eliminate all remaining industrial waste-water discharges to streams, enforcing a "zero emissions" policy for the waste water from our factories.

20 Upgrade water treatment plants so they can filter out chemicals and pharmaceuticals. Most plants do not handle either.

21 Buy organic food.

22 Only use pharmaceuticals when absolutely necessary. Learn about natural cures and how important good nutrition, sleep, and low stress levels are to keep you healthy – and pharma-free.

23 If you have to dispose of old paint, varnish, or other DIY chemicals, check with your local government's environment or public works office to find out the safest way to do so.

24 Today's society as a whole has reached unparalleled levels of prosperity. Water pollution solutions ARE affordable and practical – and essential to our future well-being. Let's get on with it!

---

inundation *n.* 洪水；（洪水般）布满
reinvigorate *vt.* 使重新振作
foot the bill 付账
pharmaceutical *n.* 医药品

Unit 4

Soil pollution

In this unit, you will learn:

- **Subject-related knowledge:** Definition, causes and effects of soil pollution
  Solutions to soil pollution and erosion
- **Academic skill:** Writing an abstract
- **Reading strategy:** Identifying the structure of a complex sentence

Section A

Pre-reading

1 Match the expressions with the pictures below.

deforestation   pesticide and fertilizer   soil salination
waterlogging   acid rain   indiscriminate dumping of trash

1. _____   2. _____   3. _____

4. _____   5. _____   6. _____

2 Discuss the following questions with your partner

1. What do you think is soil pollution? Why does it occur?
2. What can we contribute to reducing soil pollution, locally and globally?

Text A

What is soil pollution?

1. With the rise of concrete buildings and roads, one part of the Earth that we rarely see is the soil. It has many different names, such as dirt, mud, and ground. Keeping it healthy is essential to maintaining a beautiful planet. However, like all other forms of nature, soil also suffers from pollution.

2. Soil pollution or soil contamination refers to anything that causes contamination of soil and degrades the soil quality. It occurs when the pollutants causing the pollution reduce the quality of the soil and convert the soil inhabitable for microorganisms and macroorganisms living in the soil. Soil pollution occurs when the presence of toxic chemicals, pollutants or contaminants in the soil is in high enough concentrations to be of risk to plants, wildlife, humans and of course, the soil itself.

3. The loss of good soil that sustains human beings is an increasingly serious issue today. The United Nations Food and Agriculture Organization states that annually, 75 billion tons of soil, the equivalent of nearly 10 million hectares,

which is about 25 million acres, of arable land is lost to erosion, water-logging and salination and another 20 million hectares is abandoned because its soil quality has been degraded. Contact with contaminated soil may be direct, from using parks, schools etc., or indirect by inhaling soil contaminants which have vaporized or through the consumption of plants or animals that have accumulated large amounts of soil pollutants, and may also result from secondary contamination of water supplies and from deposition of air contaminants.

4 Soil pollution can occur either because of human activities or because of natural processes. However, mostly it is due to human activities such as mining, modern practices in agriculture, indiscriminate dumping of trash and unregulated disposal of untreated wastes of various industries.

1. Industrial activities

5 Industrial activities have been the biggest contributor to the problem in the last century, especially since the amount of mining and manufacturing has increased. Most industries are dependent on extracting minerals from the Earth. Whether it is iron ore or coal, the by-products may be contaminated, and usually they are not disposed of in a manner that can be considered safe. As a result, the industrial waste lingers in the soil surface for a long time and makes it unsuitable for use.

2. Agricultural activities

6 Chemical utilization has gone up tremendously since technology provided us with modern pesticides and fertilizers. They are full of chemicals that are not produced by nature and cannot be broken down by it. As a result, they seep into the ground after they mix with water and slowly reduce the fertility of the soil.

7 Other chemicals damage the composition of the soil and make it easier to erode by water and air. Plants absorb many of these pesticides and when they decompose, they cause soil pollution since they become a part of the land.

3. Waste disposal

8 A growing cause for concern is how we dispose of our waste. While industrial waste is sure to cause contamination, there is another way in which we are adding to the pollution. Every human produces a certain amount of personal waste products in the form of urine and feces. While much of it moves into the sewer system, there is also a large amount that is dumped directly into landfills in the forms of diapers. Even the sewer system ends at the landfill, where the biological waste pollutes the soil and water. This is because our bodies are carrying a certain amount of toxins and chemicals which are now seeping into the land and causing pollution of soil.

4. Deforestation

9 Because of deforestation and forest fires, soils lose their vegetation cover. The erosion process is thus accelerated, creating soil degradation as well as water pollution. Deforestation leads to the loss of the land's value as, once converted into a dry or barren land, it may never be made fertile again.

5. Acid rain

10 Acid rain is caused when pollutants present in the air mix up with the rain and fall back on the ground. The polluted water could dissolve away some of the important nutrients found in soil and change the structure of the soil.

11 Since soil pollution is not a lone standing entity, its effects are carried over as water pollution and air pollution. It affects every aspect of the environment and every organism from the earthworm to humans. Some of the adverse effects are as follows:

- **Effect on human health**

12 Considering how soil is the reason we are able to sustain ourselves, the contamination of it has major consequences on our health. Crops and plants that are grown on polluted soil absorb much of the pollution and then pass it on

to us. This could explain the sudden surge in small and terminal illnesses. For example, living, playing or working on polluted soil can cause skin complaints, respiratory issues and other ailments.

13 Long-term exposure to such soil can affect the genetic make-up of the body, causing congenital illnesses and chronic health problems that cannot be cured easily.

• Effect on growth of plants

14 The ecological balance of any system gets affected due to the widespread contamination of the soil. Most plants are unable to adapt when the chemistry of the soil changes so radically in a short period of time. Fungi and bacteria found in the soil that bind it together begin to decline, which creates an additional problem of soil erosion.

15 The toxic chemicals present in the soil can decrease soil fertility and therefore make land unsuitable for agriculture and any local vegetation to survive. The soil pollution causes large tracts of land to become hazardous to health. Unlike deserts, which are suitable for its native vegetation, such land cannot support most forms of life.

• Effect on ecosystem and biodiversity

16 It is relatively easy to undertake a cost-benefit analysis of the use of pesticides, fungicides, and other plant protection products as they relate directly to food production outcomes. It is much more difficult, however, to estimate the externalities associated with their use, particularly for major ecosystem processes such as soil biodiversity or water quality, and for human health.

17 For instance, filtering and buffering of substances in water, and the transformation of contaminants, is one of the essential regulating services provided by soils. Soils are also alive with microorganisms, bacteria, and micro-fauna which are responsible for much of the nutrient cycling processes

essential for food production and terrestrial carbon stocks — and we know relatively little about long-term exposure (or mixture) effects of agrochemicals on soil biodiversity or people.

18 A number of ways have been suggested to curb the current rate of soil pollution. Such attempts at cleaning up the environment require plenty of time and resources to be pitched in. Industries have been given regulations for the disposal of hazardous waste, which aims at minimizing the area that becomes polluted. Organic methods of farming are being supported, which do not use chemical-laden pesticides and fertilizers. Use of plants that can remove the pollutants from the soil is being encouraged. However, the road ahead is quite long and the prevention of soil pollution will take many more years.

New words and expressions

contamination /kənˌtæmɪˈneɪʃən/ n.
the state of being polluted 污染；玷污

degrade /dɪˈgreɪd/ vt.
to make sth. become worse, especially in quality 使…降低；使…恶化

equivalent /ɪˈkwɪvələnt/ n.
a person or thing equal to another in value, measure, force, effect, significance, etc. 对等的人（或事物）

arable /ˈærəbl/ adj.
(of farmland) capable of being farmed productively 适于耕种的

erosion /ɪˈrəʊʒən/ n.
condition in which the earth's surface is worn away by the action of water and wind 侵蚀；腐蚀

waterlogging /ˈwɔːtəˌlɒgɪŋ/ n.
water soaking 水浸；渍涝

salination /ˌsælɪˈneɪʃən/ n.
the process of increasing the salt content in the soil 盐碱化

deposition /ˌdepəˈzɪʃən/ n.
the natural process of laying down a deposit of sth. 沉积；沉淀

indiscriminate /ˌɪndɪsˈkrɪmɪnət/ adj.
failing to make or recognize distinctions 任意的；无差别的

fertility /fəˈtɪləti/ n.
the property of producing abundantly and sustaining vigorous and luxuriant growth 多产；肥沃

decompose /ˌdiːkəmˈpəʊz/ vi.
to separate (substances) into constituent elements or parts 分解

urine /ˈjʊərɪn/ n. 尿液

feces /ˈfiːsiːz/ n. 排泄物；粪便

landfill /ˈlændfɪl/ n.
a large deep hole in which very large amounts of rubbish are buried 垃圾填埋场

toxin /ˈtɒksɪn/ n. 毒素；毒质

deforestation /ˌdiːˌfɔːrɪˈsteɪʃən/ n.
the state of being clear of trees 砍伐森林

dissolve /dɪˈzɒlv/ vt.
to cause to go into a solution 使…溶解；使…液化

terminal /ˈtɜːmɪnəl/ adj.
a terminal illness cannot be cured and will cause someone to die, usually slowly （疾病）不治的，晚期的

respiratory /ˈrespərətəri/ adj.
relating to breathing 呼吸的

ailment /ˈeɪlmənt/ n.
an illness that is not very serious 小病

congenital /kənˈdʒenɪtəl/ adj.
present at birth but not necessarily hereditary 先天的；天生的

chronic /ˈkrɒnɪk/ adj.
a chronic illness or pain is serious and lasts for a long time （疾病）慢性的；（疼痛）长期的

fungi /ˈfʌndʒaɪ/ n. (plural form of fungus /ˈfʌŋgəs/) 真菌

bacteria /bækˈtɪəriə/ n.
(plural form of bacterium /bækˈtɪəriəm/) 细菌

externality /ˌekstɜːˈnæləti/ n.
the quality or state of being outside or directed toward or relating to the outside or exterior 外在性；外部事物

iron ore /ˈɔː(r)/ n.
an ore from which iron can be extracted 铁矿石；铁矿砂

micro-fauna /ˌmaɪkrəʊˈfɔːnə/ n. 微动物群

terrestrial /tə'restriəl/ *adj.*
living on or relating to land rather than water 陆地的

chemical-laden /'kemɪkəl-'leɪdən/ *adj.*
filled with a great quantity of chemicals 充满化学品的

Reading comprehension

1 Complete the following table based on the information from the text.

Section	Content
Introduction (Paras.1-3)	Soil pollution refers to anything that causes 1) _____ of soil and 2) _____ the soil quality. The loss of good soil that sustains human beings is increasingly a serious issue today.
3) _____ of soil pollution (Paras.4-11)	4) _____ activities 5) _____ activities 6) _____ disposal 7) _____ 8) _____ rain
Adverse 9) _____ of soil pollution (Paras.12-18)	Effect on 10) _____ Effect on 11) _____ Effect on 12) _____ and 13) _____
Conclusion (Paras.19)	A number of ways have been suggested to prevent soil pollution. For example: 14) _____ 15) _____ 16) _____

Language focus

1 Match the specialized words or expressions in Column A with their definitions in Column B. Then fill in the blanks in the sentences with these specialized words.

Column A	Column B
___ 1. congenital illness	A. a process used to measure the benefits of a decision or taking action minus the costs associated with making that decision or taking that action
___ 2. fungicide	B. rain polluted by acid that has been released into the atmosphere from factories and other industrial processes
___ 3. cost-benefit analysis	C. a disease that a person has had from birth
___ 4. micro-fauna	D. a chemical that can be used to kill fungus or to prevent it from growing
___ 5. skin complaint	E. the amount of a substance in a liquid or in another substance
___ 6. salination	F. microscopic animals, especially those inhabiting the soil, an organ, or other localized habitat
___ 7. acid rain	G. genetic illnesses or infections causing skin to react in unusual ways
___ 8. concentration	H. a build-up of salts in soil, eventually to toxic levels for plants

1. A(n) _____ is a procedure which takes into account all costs and all benefits of a project.
2. Eczema (湿疹) is a common _____ which often runs in families.
3. It is well known that Helen Keller's deafness is not a(n) _____.
4. More steel mills and chemical plants mean more _____ and smog, not to mention global warming.
5. Human's irrational development activities have brought about serious effects on soil environment, such as desertification, soil erosion, _____, and contamination, etc.

6. The latest data showed that global ozone _____ had dropped several per cent over the last decade.
7. This company is a high-tech chemical enterprise, which professionally manufactures herbicide, pesticide and _____.
8. The groups of soil _____ responded differently to soil environmental factors.

2 Fill in the blanks with the words given below. Change the form where necessary.

accumulate   composition   complaint   arable   contaminate
exposure   equivalent   chronic   terminal   indiscriminate

1. The research also suggested taking measures to reduce _____ to chemicals in our environment.
2. In this study, the acute and _____ effects of earthworms on soil erosion were evaluated across different endpoints.
3. Weathering of the Earth's crusts by different processes leads to the formation of soil that _____ over the centuries.
4. The _____ use of fertilizers is damaging the environment.
5. The leaking of sewerage system can affect soil quality and cause soil pollution by changing the chemical _____ of the soil.
6. Think of life as a(n) _____ illness. Because if you do, you will live it with joy and passion, as it ought to be lived.
7. Excessive use of inorganic nitrogen fertilizers leads to acidification of soil and _____ the agricultural soil.
8. According to the survey conducted by the United Nations Food and Agriculture Organization, a(n) _____ of one soccer pitch of soil is eroded every five seconds.
9. The detrimental effects of soil pollution are widespread and many. For example, it can cause skin _____ like dryness, premature aging, skin rashes, eczema and acne.
10. As well as the noise, traffic, loss of _____ land and threat to birds, airports are a major source of pollution and greenhouse gas emissions.

3 Study the affixes / suffixes and their meanings below with the examples from Text A. Then try to add the affixes / suffixes to the given words in the brackets to finish the sentences with the newly formed words. Change the form if necessary.

Affixes	Meaning	Examples
bio-	relating to or using living things	biodiversity
micro-	extremely small	microorganism
de-	opposite	degrade; deforestation; decompose; deposition
dis-	opposite	dispose; discontent
in-	opposite	inhabitable; indiscriminate
un-	opposite	unregulated; untreated
-cide	indicating a person or thing that kills	pesticide; fungicide
-able	describing sth. that can be done	arable; inhabitable

1. He's looking forward to seeing results from several ongoing studies that are evaluating new _____ methods for reducing soil pollution. (technical)
2. The ecological risk of contaminated soil, _____ application, sewage sludge amendment, and other human activities will be studied in this project. (pest)
3. There are abundant _____ and nutrition in sludge which can increase crop yields and improve crop qualities. (organism)
4. The government decided to plant trees on the _____ land and change it into a forest. (grade)
5. Soil pollution, though _____, may be the most difficult to deal with. (visible)
6. Rain in this area is slight acidic even in _____ air, because carbon dioxide in the atmosphere and other natural acid-forming gases dissolve in the water. (pollute)

7. Since the metals are not _____, their accumulation in the soil above their toxic levels becomes an indestructible (不可破坏的) poison for crops. (degrade)
8. Faced with public _____, this country has started to address air pollution by modernizing factories and moving coal-consuming industries away from cities. (content)
9. Pesticides which are not rapidly _____ may exert harmful effects on microorganisms, as a result of which plant growth may be affected. (compose)
10. Because of increasing pollutants, the water in this lake has been _____ now. (drink)

4 Translate the following paragraph into English.

土壤污染是指土壤中含有高浓度的有毒污染物，从而对人类健康和生态系统构成威胁的现象。造成土壤污染的主要原因是大众缺乏环境意识。例如，人类过度使用杀虫剂，致使土壤失去肥力。为有效预防和减少土壤污染，大家都要响应"减少、再利用和回收"的环保宣言，一起行动起来(pitch in)。

Critical thinking

1 Work in groups and brainstorm steps you can take as young citizens to help reduce soil pollution in your community.

2 December 5th is celebrated as World Soil Day. Please form a group of four or five students to design a vivid and clearly worded poster slogan for celebrating and publicizing this year's World Soil Day. (For example: Caring for the planet starts from the ground!)

Research task

Academic skill: Writing an abstract

An abstract is a summary of a research article, through which readers could easily identify the main content of the article so as to determine whether it is relevant or useful to their research. It is usually put at the beginning of the research article.
To write a successful abstract, you are supposed to follow some basic principles in terms of content and language.

1. Content

   An abstract often consists of four parts: purpose, method, result and conclusion. The following table provides some sentence structures that can be used in the four parts.

Part of an abstract	Possible sentence structure
Purpose	- This paper seeks to justify ... in terms of ... - The test is aimed to ... - The study is intended to ...
Method	- The formula is derived from ... - The test is carried out by using ... - The analysis is made with ...
Result	- It is indicated that ... - It is found / observed that ... - The results are tabulated to show that ...
Conclusion	- It is concluded that ... - This paper concludes that ...

2. Language

   Length

   Abstracts could be within 200-300 words or longer than this. And the abstract should also limit it to new information, which means that literature review is not included in an abstract.

Tense

Generally, there are three tenses used frequently in an abstract, namely, the simple present tense, the simple past tense and the present perfect tense, which are elaborated in the following table.

Tense	When to use in an abstract
Simple present tense	<ul><li>Describing background information.</li><li>Stating a fact or theory that is widely accepted.</li><li>Stating conclusions.</li></ul>
Simple past tense	<ul><li>Describing the research method.</li><li>Describing the research activity.</li><li>Reporting results.</li></ul>
Present perfect tense	<ul><li>Referring to a previous study with results that are still relevant.</li><li>Emphasizing the influence or significance of the research.</li></ul>

Voice

Both active and passive voices are adopted in an abstract, but in different parts. Since research articles tend to describe and explain research results in an objective way, passive voice, compared with active voice, is more frequently used. However, when introducing research purpose and conclusion, active voice is used more often.

Sample

Abstract: Urban gardening is popular in many cities. However, many urban soils are contaminated and pose risks to human health. This study was conducted in a highly publicized urban garden in Brooklyn, NY with elevated Pb and As levels. Our objectives were to: (1) assess the nature and extent of Pb and As contamination at this site; (2) evaluate the effectiveness of amendments on reducing the bioaccessibility and phytoavailability of Pb and As in soil; and (3) assess the potential exposure of children to Pb and As through direct and indirect exposure pathways. Field surveys of the site revealed that contamination was highly concentrated in one area of the garden associated with fruit tree production. Field plots were established in this area, with three different treatments (bone meal, compost, sulfur) and an unamended control. Bioaccessibility of Pb was significantly reduced by all three treatments compared to the control (33%): bone meal (24%), compost (23%), sulfur (24%). In this study, As bioaccessibility remained high (80-93%) with or without treatments. We found that the effectiveness of soil remediation with amendments is variable and often limited, and contaminated sites can still pose a significant risk to urban gardeners. The results of a simple assessment model suggested that Pb and As exposure was mostly from soil and dust ingestion, rather than vegetable consumption. This work is unique in that it evaluates actual elevated levels of contamination, in actively gardened urban soils, in a highly visible public context. It fills important gaps between basic research and analysis of human exposure to toxic trace metals that can be a constraint on a highly beneficial activity.

Task

Work in groups and search an academic paper on soil pollution and analyze its abstract. If you do not think the abstract is typical, please polish it according to the Academic skill above.

Section B

Reading strategy

Identifying the structure of a complex sentence

Sentences in formal technical texts usually contain lots of information and turn out to be very complex with many layers of structure. For example:

*Contact with contaminated soil may be direct, from using parks, schools, etc., or indirect by inhaling soil contaminants **which** have vaporized or through the consumption of plants or animals **that** have accumulated large amounts of soil pollutants.*

To identify the framework of this sentence, you should:

1) Identify the basic frame of the complex sentence.
2) Read through the sentence, and target the linking words: that and which.
3) Decide the function of the linking words.

For example, in this sentence, the basic frame is "Contact with contaminated soil may be direct from …, or indirect by …" Then we can target two linking words: "which" and "that", introducing two attributive clauses "which have vaporized or through the consumption of plants or animals that have accumulated large amounts of soil pollutants" and "that have accumulated large amounts of soil pollutants". The former attributive clause modifies "soil contaminants". The latter attributive clause modifies "plants or animals".

Task

Now read Text B and apply the strategy above to identify the structure of the underlined sentences.

Text B

Reducing soil pollution and erosion

1   Arable land is turning to desert and becoming non-arable at ever-increasing rates, due to global warming and agricultural fertilizers and pesticides, lessening the hope that we can feed our booming population. And world population is still growing. Food production will have to increase at least 40% and most of that will have to be grown on the fertile soils that cover just 11% of the global land surface. Actually, there is little new land that can be brought into production and existing land is being lost and degraded. However, we can make a difference to help purify soils and restore balance to once fertile grounds, including your own backyard or farm.

2   Reduce deforestation and begin reforestation–Deforestation and soil erosion are very much interconnected. For example, the effects of acid rain and floods

can decimate healthy soil in the absence of trees, which would otherwise help absorb and maintain these waters and the toxins that come along. Soil erosion can occur when there are no trees or few plants to prevent the top layer of soil from being removed and transported by forces of nature, such as water and air, which contribute to soil pollution. Through reforestation efforts and planting new vegetation in areas that are prone to erosion, soil pollution can be further prevented.

3 **Avoid intensive farming practices,** such as overcropping and overgrazing, as it leads to flood and soil erosion and further deterioration of the soil layer. For example, controlled grazing or management-intensive grazing (also known as rotational grazing) can be adopted to check unlimited access of animals to pastures and also to manage the grazing land effectively.

4 **Reduce your "waste footprint"** — Waste, such as plastic, non-biodegradable materials and litter, can accumulate in fertile land, polluting and altering the chemical and biological properties of soil. According to the Clean Air Council, almost one-third of the waste in the U.S. comes from packaging — try to purchase materials with the least amount of packaging and always Reduce, Reuse and Recycle!

5 **Discover soil washing,** which uses water to remove contaminants from soils by "scrubbing" soil to remove and separate the portion of the soil that is most polluted. Soil washing reduces the amount of soil needing further cleanup and is typically used along with other methods to clean up the soil as it is usually not sufficient enough to do the job alone. Soil washing allows the cleanup of polluted soil in place without having to excavate.

decimate *vt.* 毁坏；杀死
be prone to 易于⋯，有⋯倾向
overgrazing *n.* 过度放牧
deterioration *n.* 恶化，退化
scrub *vt.* 刷洗
excavate *v.* 挖掘

6. **Discover bioremediation** — Use and incite the growth of naturally-occurring microorganisms to break down contaminants and remediate soil pollution by using them as a food source during the aerobic processes, which requires the right temperature, nutrients and amount of oxygen in the soil.

7. **Maintain a vegetation cover,** especially in the vulnerable areas, such as steep slopes and arid areas. Many studies have shown that in a wide range of environments, both runoff and sediment loss decrease exponentially with an increasing percentage of vegetation cover, also known as plant or tree cover. Forest vegetation significantly reduces summer soil surface temperatures and a vegetation cover of 45%-50% effectively protects soils from raindrop impact and significantly reduces soil erosion.

8. **Learn to compost** — Composting is nature's process of recycling decomposed organic materials into rich soil known as compost. Composting commonly comes in the form of vegetable compost and animal manure compost. Mix in earthworms to help remediate the waste, remove toxic heavy metals, break down the organic matter and leave richer soil. Not sure where to buy composts? Keep your kitchen scraps from vegetables, fruits, coffee grounds, etc. in a composting bin or container. You can use this to add a compost site to your garden or add to your garden to create deeper topsoil, recycle nutrients, and save landfill space and reduce water and soil pollution.

9. **Use crop rotation in cultivation areas,** according to Pesticide Action Center Europe, to prevent the soil from becoming exhausted to help create a biodiverse soil, which reduces the chance of soil bound organisms to get a pest, reduces

---

incite *vt.* 刺激；引起
remediate *vt.* 补救；修复；革除
aerobic *adj.* 需氧的，好氧的
exponentially *adv.* 成指数地，成倍地
compost *vi.* 堆制成堆肥 *n.* 堆肥
manure *n.* 动物粪肥
rotation *n.* 轮作，轮种

the use of pesticides, helps minimize the growth of weeds and helps form a good soil structure that can result in a higher yield. Crop rotation performed with nitrogen-fixating rotation crops will reduce the input of fertilizers and, consequently, the pollution by nitrogen. Crop rotation combined with conservation tillage will lead to higher soil-carbon content, which helps combat climate change and soil erosion.

10. **Reduce the use of chemical pesticides, fertilizers, insecticides and weedicides,** whose overuse leads to soil pollution. These harsh chemicals may kill the intended weeds or insects that can damage and stunt the proper growth of the plant, but they are not easily broken down. Instead, they accumulate in the soil, and can be responsible for depleting the fertility of the soil and be a threat not only to the plants, but also to the animals and humans consuming it. When these chemicals are used in the soil, the non-degradable metals that accompany the essential nutrients slowly begin to amass in the soil above preferred toxic levels, such as the notorious, damaging phosphate fertilizers. To give plants the nutrients they need for growth, while protecting the health and longevity of the soil, you should choose low-phosphorus fertilizers, try making your own organic fertilizer using a compost bin, or simply purchase organic fertilizers, which are harmless to nature and nourish the soil, unlike chemical fertilizers. Try to avoid pesticides as much as possible. Instead, try using organic pesticides, which help control pests without damaging the environment.

11. **Purchase certified organic products,** which must be produced without the use of synthetic fertilizers, along with many other benefits.

12. **Eliminate weeds,** which absorb the necessary minerals out of the soil. Eliminating weed growth can decrease soil pollution to a great extent. A common method of reducing weed growth is to prevent light from reaching the

---

tillage *n.* 耕作，耕地
phosphate *n.* 磷酸盐（酯）

weeds in order to kill them by covering the soil with numerous layers of wet newspapers or a plastic sheet for several weeks before cultivation.

13 With a global population that is projected to exceed 9 billion by 2050, our current and future food security hinges on our ability to increase yields and food quality using the soils that we already have available today. Soil pollution negatively impacts us all, and has been identified as one of the main threats to soil functions worldwide.

14 We need to be aware of the causes of soil pollution so that we can create and implement solutions. Soil protection and conservation starts with us. Making sustainable food choices, properly recycling dangerous materials like batteries, composting at home to reduce the amount of waste that enters landfills or managing antibiotic waste more responsibly, are just a few examples of how we can be part of the solution. On a larger scale, we need to promote sustainable agricultural practices in our communities.

15 Healthy soil is a precious, non-renewable commodity that is increasingly threatened by destructive human behaviors. We are responsible for the soils that provide us with food, water and air, and we need to take action today to ensure that we have healthy soils for a sustainable and food-secure future.

project *vt.* 预测
hinge on 取决于；以…为转移

Unit 5

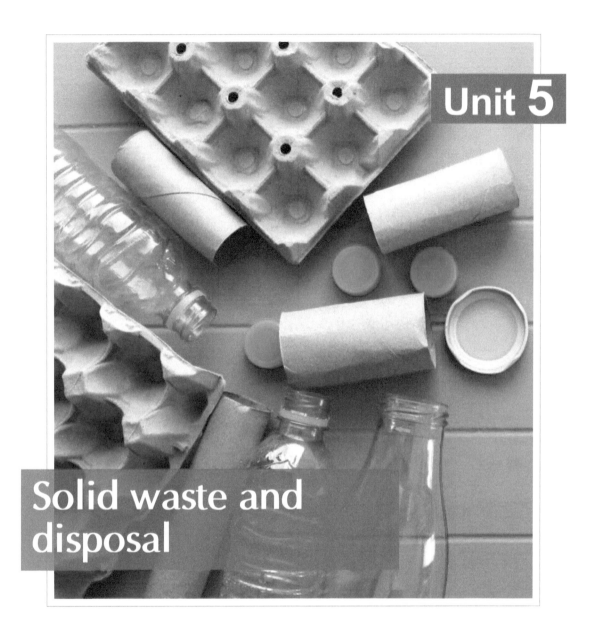

Solid waste and disposal

In this unit, you will learn:

- **Subject-related knowledge:** Solid waste disposal in the U.S.
  Better solid waste disposal
- **Academic skill:** Structure of academic papers
  Using academic language
- **Reading strategy:** Outlining (Part I)

Section A

Pre-reading

Look at the pictures below and answer the following questions.

1. Recognize the means of waste disposal in the above pictures and tell how you usually handle waste.
2. Give an example of a community or country that does well in waste disposal and discuss what you can learn from them.

1. The RCRA (The Resource Conservation and Recovery Act) of the United States defines the term "solid waste" as any garbage (e.g. milk cartons and coffee grounds), refuse (e.g. metal scrap, wall board, and empty containers), sludge from waste treatment plants, water supply treatment plants, or pollution control facilities (e.g. scrubber slags), industrial waste (e.g. manufacturing process wastewaters and nonwastewater sludge and solids), and other discarded materials, including solid, semi-solid, liquid, or contained gaseous materials resulting from industrial, commercial, mining, agricultural, and community activities (e.g. boiler slags).

2. The definition of "solid waste" is not limited to waste that is physically solid. Many kinds of solid waste are liquid, while others are semi-solid or gaseous.

3. The term "solid waste" is very broad, including not only the traditional nonhazardous solid wastes, such as municipal garbage and industrial wastes, but also hazardous wastes. Hazardous waste, a subset of solid waste, is regulated under RCRA Subtitle C. For purposes of regulating hazardous wastes, EPA (the United States Environmental

Text A

Solid waste disposal in the U.S.

Protection Agency) established by regulation a separate definition of "solid waste". (Hazardous waste is not to be discussed here.)

4 Nonhazardous solid wastes mainly include municipal solid waste (MSW) and industrial waste.

Municipal solid waste

5 Municipal solid waste is a subset of solid waste and is defined as durable goods (e.g. appliances, tires, batteries), nondurable goods (e.g. newspapers, books, magazines), containers and packaging, food wastes, yard trimmings, and miscellaneous organic wastes from residential, commercial, and industrial nonprocess sources.

6 To address the increasing quantities of municipal solid waste, EPA recommends that communities adopt "integrated waste management" systems tailored to meet their needs. The term "integrated waste management" refers to the complementary use of a variety of waste management practices to safely

and effectively handle the municipal solid waste stream. An integrated waste management system will contain some or all of the following elements: source reduction, recycling (including composting), waste combustion for energy recovery, and / or disposal by landfilling.

7 **Source reduction:** Rather than managing waste after it is generated, source reduction changes the way products are made and used in order to decrease waste generation. Source reduction, also called waste prevention, is defined as the design, manufacture, and use of products in a way that reduces the quantity and toxicity of waste produced when the products reach the end of their useful lives. The ultimate goal of source reduction is to decrease the amount and the toxicity of waste generated. Businesses can manufacture products with packaging that is reduced in both volume and toxicity. They also can reduce waste by altering their business practices (e.g. reusing packaging for shipping, making double-sided copies, maintaining equipment to extend its useful life, and using reusable envelopes). Community residents can help reduce waste by leaving grass clippings on the lawn or composting them with other yard trimmings in their backyards, instead of bagging such materials for eventual disposal. Consumers play a crucial role in an effective source reduction program by purchasing products having reduced packaging or that contain reduced amounts of toxic constituents. This purchasing subsequently increases the demand for products with these attributes.

8 **Recycling:** Municipal solid waste recycling refers to the separation and collection of wastes, their subsequent transformation or remanufacture into usable or marketable products or materials, and the purchase of products made from recyclable materials.

9 Communities can offer a wide range of recycling programs to their businesses and residents, such as drop-off centers, curbside collection, and centralized composting of yard and food wastes.

10 The U.S. government has developed several initiatives in order to bolster the

use of recycled products. EPA's federal procurement guidelines are designed to bolster the market for products manufactured from recycled materials. The procurement program uses government purchasing to spur recycling and markets for recovered materials.

11 **Combustion:** Confined and controlled burning, known as combustion, can not only decrease the volume of solid waste destined for landfills, but also recover energy from the waste-burning process.

12 There are three types of technologies for the combustion of MSW: mass burn facilities, modular systems, and refuse derived fuel systems. Mass burn facilities are by far the most common types of combustion facilities in the U.S. The waste used to fuel the mass burn facilities may or may not be sorted before it enters the combustion chamber. Many advanced municipalities separate the waste on the front end to pull off as many recyclable products as possible. Modular systems are designed to burn unprocessed, mixed MSW. They differ from mass burn facilities in that they are much smaller and are portable and can be moved from site to site. Refuse derived fuel systems use mechanical methods to shred incoming MSW, separate out non-combustible materials, and produce a combustible mixture suitable as a fuel in a dedicated furnace or as a supplemental fuel in a conventional boiler system.

13 **Landfilling:** Landfilling of solid waste still remains the most widely used waste management method. Many communities in the U.S. are now having difficulties siting new landfills, largely as a result of increased citizen concerns about the potential risks and aesthetics associated with having a landfill in their neighborhood. To reduce risks to health and the environment, EPA developed minimum criteria that solid waste landfills must meet.

Industrial waste

14 Industrial waste is also a subset of solid waste and is defined as solid waste generated by manufacturing or industrial processes that is not hazardous

waste. Such waste may include, but is not limited to, waste resulting from the following manufacturing processes: electric power generation; fertilizer or agricultural chemicals; food and related products or by-products; inorganic chemicals; iron and steel manufacturing; leather and leather products; nonferrous metals manufacturing or foundries; organic chemicals; plastics and resins manufacturing; pulp and paper industry; rubber and miscellaneous plastic products; stone, glass, clay, and concrete products; textile manufacturing; transportation equipment; and water treatment. Industrial waste does not include mining waste or oil and gas production waste.

15 Similarly to municipal solid waste, EPA recommends considering pollution prevention options when designing an industrial waste management system. Pollution prevention will reduce waste disposal needs and can minimize impacts across all environmental media. Pollution prevention can also reduce the volume and toxicity of waste. Lastly, pollution prevention can ease some of the burdens, risks, and liabilities of waste management. As with municipal solid waste, EPA recommends a hierarchical approach to industrial waste management: First, prevent or reduce waste at the point of generation (source reduction); second, recycle or reuse waste materials; third, treat waste; and finally, dispose of remaining waste in an environmentally protective manner.

16 **Source reduction:** Source reduction activities for industrial waste include equipment or technology modifications; process or procedure modifications; reformulations or redesign of products; substitution of less-noxious product materials; and improvements in housekeeping, maintenance, training, or inventory control.

17 **Recycling:** Industry can benefit from recycling: the separation and collection of byproduct materials, their subsequent transformation or remanufacture into usable or marketable products or materials, and the purchase of products made from recyclable materials. Many local governments and states have established materials exchange programs to facilitate transactions between generators of byproduct materials and industries that can recycle wastes as raw materials.

Materials exchanges are an effective and inexpensive way to find new users and uses for a byproduct material.

18 **Treatment prior to disposal:** Treatment of nonhazardous industrial waste is not a federal requirement. However, it can help to reduce the volume and toxicity of waste prior to disposal. Treatment can also make waste amenable for reuse or recycling. Consequently, a facility managing nonhazardous industrial waste might elect to apply treatment.

Landfilling

19 As with municipal solid waste, industrial facilities will not be able to manage all of their industrial waste by source reduction, recycling, and treatment. Landfilling is the least desirable option and should be implemented as part of a comprehensive waste management system. Industrial waste landfills can face opposition as a result of concerns about possible negative aesthetic impact and potential health risks. To reduce risks to health and the environment, EPA also developed minimum criteria for industrial waste landfills to meet.

New words and expressions

sludge /slʌdʒ/ *n.*
a thick soft substance that remains when liquid has been removed from sth. in an industrial process（工业过程中去除液体后的）烂泥状混合物

nonhazardous /ˌnɒnˈhæzədəs/ *adj.*
unlikely to harm people's health 无害的

municipal /mjuːˈnɪsɪpəl/ *adj.*
connected with or belonging to a town, city or district that has its own local government 市政的；都市的

subset /ˈsʌbset/ *n.*
a smaller group of people or things formed from the members of a larger group 子集

trimmings /ˈtrɪmɪŋz/ *n.*
(pl.) 修剪下来的东西

miscellaneous /ˌmɪsəˈleɪniəs/ *adj.*
consisting of many different kinds of things that are not connected and do not easily form a group 混杂的；各种各样的

complementary /ˌkɒmplɪˈmentəri/ *adj.*
different but together form a useful or attractive combination of skills, qualities or physical features 互补的；补充的；相互补足的

toxicity /tɒkˈsɪsəti/ *n.*
the quality of being poisonous; the extent to which sth. is poisonous 毒性；毒力

constituent /kənˈstɪtjuənt/ *n.*
one of the parts of sth. that combine to form the whole 成分；构成要素

modular /ˈmɒdjʊlə(r)/ *adj.*
based on modules or made using modules 分单元的；模块化的；组合式的，组件式的

site /saɪt/ *vt.* 场所；地址

aesthetics /iːsˈθetɪks/ *n.* 美感；审美观

nonferrous /nɒnˈferəs/ *adj.*
denoting any metal other than iron 不含铁的；非铁的

foundry /ˈfaʊndri/ *n.*
a factory where metal or glass is melted and made into different shapes or objects 铸造厂；玻璃厂

resin /ˈrezɪn/ *n.*
a sticky substance that is produced by some trees and is used in making varnish, medicine, etc. 树脂

liability /ˌlaɪəˈbɪləti/ *n.*
the state of being legally responsible for sth.（法律上对某事物的）责任，义务

hierarchical /ˌhaɪəˈrɑːkɪkəl/ *adj.*
classified according to various criteria into successive levels or layers 按等级划分的；等级（制度）的

inventory /ˈɪnvəntəri/ *n.*
the goods in a shop 存货；库存

amenable /əˈmiːnəbl/ *adj.*
(fml.) capable of being treated or dealt with in a particular way 可解决的

implement /ˈɪmplɪmənt/ *vt.*
to make sth. that has been officially decided start to happen or be used 实施；执行

solid waste 固体废料；固体垃圾

coffee grounds 咖啡渣

metal scrap /skræp/ 废金属

scrubber slag /ˈskrʌbə(r) slæg/ 洗刷物残渣

source reduction 源头削减

Reading comprehension

Categorize the following wastes based on the information from Text A.

A. leather
B. newspaper
C. stone, glass, clay, and concrete
D. yard trimmings
E. banana peel
F. plastics and resins

Municipal waste	Industrial waste

Language focus

1 Complete the following sentences by translating the Chinese in brackets into English.

1. _____ (庭院修整废物) can be turned into a soil amendment for crop production.
2. The benefits of _____ (源头削减) greatly outweigh those of incineration and landfill in terms of reducing energy use, greenhouse gas emissions, and other environmental impacts.
3. The float collector can be used in the _____ (分离与收集) of floats in rivers, lakes, and the like, and thus helps find _____ (可循环材料).
4. Consumers play a crucial role in an effective source reduction program by purchasing products having reduced packaging or that contain reduced amounts of _____ (毒性成分).
5. The use of sewage _____ (烂泥状沉积物) as a fertilizer on farmland is popular in this village.

2 Match the expressions in the field of waste disposal listed in Column A with their Chinese equivalents in Column B.

Column A	Column B
___ 1. nonhazardous waste	A. 金属废料
___ 2. inventory control	B. 废物焚烧
___ 3. landfill	C. 集中堆肥
___ 4. solid waste	D. 非危险废物
___ 5. waste treatment plant	E. 库存管理
___ 6. organic waste	F. 有机废物
___ 7. centralized composting	G. 废物处理厂
___ 8. waste combustion	H. 废物填埋
___ 9. metal scrap	I. 固体废物

3 Fill in the blanks with the words given below. Change the form where necessary.

discarded   smoking   semi-solid   community   habits
garbage   benefit   discharge   treatment   regulations
preserved   recycled   belief   dangerous   hazardous

The definition of solid waste is not limited to wastes that are physically solid. Solid waste means any 1) _____ or refuse, sludge from a wastewater 2) _____ plant, water supply treatment plant, or an air pollution control facility and other 3) _____ material, including solid, liquid, 4) _____, or contained gaseous material resulting from industrial, commercial, mining, and agricultural operations, and from 5) _____ activities. Solid waste does not include solid or dissolved materials in domestic sewage, solid or dissolved materials in irrigation return flows, or industrial 6) _____

Generally, the term solid waste refers to non-hazardous waste, though according to the Resource Conservation and Recovery Act (RCRA) and other state regulations, 7) _____ waste is also a part of solid waste.

Many of the items we are throwing away can be reused, 8) _____ or composted, such as paper, glass, aluminum, metals as well as potato

and carrot peels. It is just a matter of learning new 9) _____.
Reduction, reuse, recycling and composting our trash will 10) _____
_____ all of us, our communities and our environment.

4 Translate the following paragraph into English.

在美国，未食用食物分解产生的甲烷占甲烷排放总量的23%。在从农田到餐桌的不同阶段丢弃的所有食物中，只有3%用于堆肥。绝大多数都终结于垃圾填埋场。事实上，食品是目前到达垃圾填埋场的城市固体废物中的最大组成成分。在那里，它们逐渐转化为甲烷，而甲烷是一种温室气体，它在全球变暖中的作用要比二氧化碳强25倍。

Critical thinking

1 Do some research on "Clear Your Plate" campaign and think about its effects in your university.

2 Role play a meeting in which a community activist puts forward a new waste-disposal method and different community representatives talk about the prospects and possible difficulties.

Research task

Academic skill: Structure of academic papers & using academic language

1. Structure of academic papers

   Academic papers typically have a standard structure to follow, including title, acknowledgements, abstract, introduction, literature review, research methods, results, discussion, conclusion and references, although, of course, there are variations on this basic format. For example, the reader may not find acknowledgements and may find that discussion is combined with conclusion.

   Title

   An academic paper begins with a title, which is designed to stimulate the reader's interest and inform the reader what the paper is about. Therefore, the title should be simple, and every word of it should be significant.

   Acknowledgements

   Acknowledgements provide the author an opportunity to give credit to funding bodies, departments and individuals who have been of help during their study.

   Abstract

   An abstract is a summary or a brief yet thorough overview of a paper, which informs the reader what to read in the rest part of the paper. A good abstract lets the reader know the paper is worth reading. Roughly speaking, an abstract includes background, problem, methods, results and conclusion.

   Introduction

   Most academic papers start broad and then narrow down to a specific research topic. An introduction reveals some broad knowledge and basic background information about the topic, and quickly turns to the focus of the paper. It explains the purpose and the focus of the paper in justification for the research.

   Literature review

   Literature review goes through the previous important research which relates specifically to your research topic. Also, it examines the major theories related to the topic and their contributions. It is supposed to be a synthesis of the previous literature and the new ones being researched.

Research methods

Research methods describe the research design and methodology used to complete the research. This section usually includes instruments, subjects or participants, and procedures. And remember that the reader should be provided with enough detail to replicate the research. Moreover, this section is usually written in the third person, the passive voice and the past tense.

Results

In this section, the results of the research are presented. Graphs and tables could be helpful when there is too much data to include.

Discussion

Discussion is the "heart" of the paper, in which you should: 1) restate the findings and accomplishments; 2) evaluate how the results fit in with the previous findings; 3) offer an interpretation or explanation of the results and try to ward off counter-claims; 4) list potential limitations of the study; 5) state the implications of the study and possible further research directions.

Conclusion

In this section, major claims should be reinforced in a way that is not mere summary. You need to put together all the main ideas, refer back to what you have outlined, and show the extent to which you have solved the problem.

References

References, often called a citation list, are sources consulted. This section always comes at the end of an academic paper. It must include all of the direct sources that you have referred to in the body of the paper.

2. Using academic language

What is academic language?

First, academic language is a type of formal discourse which assumes an absolutely objective tone. It is built into a relatively balanced structure and always based on scientific data. Second, it relies on glossary that emerges directly from the discourse of a given academic field. It helps to convey arguments in an authoritative tone and makes the paper trustworthy.

Use formal language

Some words are appropriate in informal situations while others are appropriate in formal or neutral contexts. In academic writing, formal or neutral language is normally preferred, e.g. using "approximately" instead of "about", "sufficient" instead of "enough". Avoid using phrasal verbs like "get up", "turn on", "wake up", and colloquialisms or idioms.

Use concise language

Academic papers should be organized in relatively short rather than long or complicated sentences, which means that no redundant information should be included. Beware of the following cases of redundancy.

Concise language	Redundant language
seemingly	It would seem that ...
an enhancement method, IDEF4x	an enhancement method which is named as IDEF4x
in landscape architecture	in the field of landscape architecture
Some woodland, in fact, has been purchased as public parks.	Some woodland, in fact, has been purchased for the purpose of creating public parks.

Use hedges

In academic writing, arguments are often stated in cautious or tentative language. This is known as "hedging". There is a wide range of words and phrases that can be used in hedging: modal auxiliary verbs (e.g. can, may, might), lexical verbs (e.g. suggest, indicate, estimate), modal adjectives (possible, likely, normal), adverbs of frequency and degree (frequently, generally, roughly, somewhat), and introductory phrases (e.g. it seems that ..., one can assume that ..., it is our view that ...). Hedging language helps to make statements as accurate and fair as possible.

Use nominalization

Nominalization is the process of transforming a verb, an adjective or a clause into a noun or noun phrase. For example, "investigate" can be nominalized into "investigation"; "a fall in hospital admissions for smoking-related diseases" is a nominalization of "fewer people have been admitted to hospitals for smoking-related diseases". Nominalization is useful in academic writing because it conveys an objective and impersonal tone.

Use signposts

Academic writing is required to be explicit. In other words, you should anticipate the reader's questions in mind. Good writers use signposting language to signal to the reader where the answers can be found. Here are examples: "the aim of this study is to ...", "in conclusion ...".

Use connectors

Connectors, or transitions, are words or phrases that allow you to move smoothly from one point or idea to the next. Here are some connectors: connectors for listing (e.g. first, firstly, second, secondly), connectors indicating addition (e.g. also, similarly), connectors indicating contrast (e.g. however, nevertheless), connectors giving a reason (e.g. for this reason, due to), connectors indicating results (e.g. as a result, consequently), connectors reformulating an idea (e.g. in other words, to put it simply), and connectors for exemplifying (e.g. for example, to exemplify).

Task

Different waste disposal methods are mentioned in Text A. Some of them are explained elaborately as in the example of source reduction. Search online for a standard academic paper on topics like landfilling, recycling, or waste combustion, etc., and evaluate its language. Ask yourself such questions as "Is the language formal or informal?", "Is the tone subjective or objective?", "Is the language concise or redundant?" and "Are the sentences and paragraphs cohesive by using the connectors?". Then write down examples found in the paper and share your evaluation in groups.

Section B

Reading strategy

Outlining (Part I)

Expository writing is very common and useful. There are many articles written in such style. It is an essential skill for students to read and understand expository writing. In order to understand the content and basic structure of the reading material, it is helpful to make an outline. It does help the reader follow the writer's line of thought, distinguish the main points from supporting details, and grasp the logical development of ideas. When you make an outline, the following steps are recommended:

1. Skim quickly to get a general overview of the content.
2. Divide the expository writing into sections. Focus on the headings and subheadings if any. (An expository writing generally can be divided into three sections: introduction, body and conclusion.)
3. Read the expository writing again, and locate and write down the thesis statement. (A writer may place the statement in the introduction, in the middle, or in the conclusion.)
4. Read paragraph by paragraph, find out the topic sentences of the supporting paragraphs, and summarize them. Make sure that what you include in your summary are key points, not minor details.
5. List the examples or details used to support the thesis statement.
6. Make a conclusion. (A conclusion may summarize the whole text, and / or recommend future action.)
7. Put what you have got into a visual organizer. Different organizational patterns may use different visual organizers.

In this unit, we will introduce three organizational patterns (see the table on the next page). Each of them has its own signal words that help decide which organizational pattern and visual organizer should be used to help make an outline.

Organizational pattern	Explanation	Visual organizer	Signal word
Order of importance	Information and ideas are arranged in order of importance.	Pyramid	first, next, finally; most important, most convincing; in addition
Cause / Effect	Details are arranged to link a result with a series of causes, or to show a logical relationship between a cause and one or more effects (e.g. describe the effect first and then explain the possible causes, or describe the cause first and then explain the effects).	Fishbone diagram	as a result of, because, due to, since, effects (result) of, cause, lead to; consequently, therefore, so, so that
Order of process	A process is described. (e.g. how to paint a color theory triangle). This type of organization can overlap substantially with a chronological organizing scheme.	Process chart	then, meanwhile, later, eventually, after a few days, next

The following three pictures are examples of each type of the visual organizers.

Pyramid

- first, first and foremost, …
- second, next, then, …
- third, besides, …
- finally, lastly, …

To reduce waste, we have to first reduce consumption, as well as minimize waste generation throughout the entire process of production, transportation and distribution of goods. Next, it is also important to change the accepted norms of consumption and strike a balance between economic activities and nature. Besides, garbage recycling helps transform waste into resources.

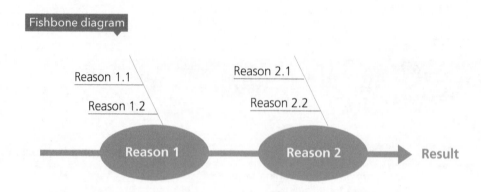

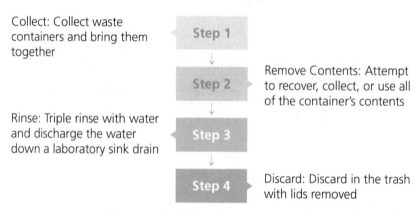

The Procedure of
Trash Disposal of Empty Containers

Collect: Collect waste containers and bring them together

Step 1

Step 2

Remove Contents: Attempt to recover, collect, or use all of the container's contents

Rinse: Triple rinse with water and discharge the water down a laboratory sink drain

Step 3

Step 4

Discard: Discard in the trash with lids removed

Task

Now read Text B and put into a process chart the changes in all those years of waste disposal methods in the town of Miyoshi where Noriko Ishizaka's family factory lies.

Better solid waste disposal

Text B

1. Disposal facilities in Japan are working hard to process the more than 380 million tons of industrial waste the country produces annually. But some communities are critical of their impact on the environment, so one woman has taken a stand by creating an innovative facility.

2. Noriko Ishizaka took over her father's company in the town of Miyoshi, about an hour's drive from Tokyo, over a decade ago. Since then, she's been working hard to overhaul waste disposal and give the industry a cleaner image.

3. The company now generates more than 40 million dollars in sales. Every day, more than 300 trucks bring in about 1,000 tons of industrial waste. All of the waste comes from demolished buildings, and contains a mix of gravel, rubble and wood, as well as steel and plastic. This type of waste is very difficult to process so it often ends up being dumped illegally.

4. "I really believe a company creates its own value when it decides to tackle waste that other companies can't handle or find unmanageable," Ishizaka says.

5. The first step in the recycling process relies on heavy

---

overhaul vt. 彻底检查；全面改革
demolish vt.（有意）拆毁（建筑物）
gravel n. 砂砾
rubble n. 碎石

machinery to separate the waste into rough categories, then materials that can be identified visually, such as wood, plastic or metal, are sorted by hand. The next step is what sets this company apart – an additional level of screening, using a range of specially developed tools. One sorting machine uses the power of suction. The waste is shaken to let paper and other light materials float to the top, where they're sucked up by a powerful vacuum. Another machine uses a magnet to catch nails and other small pieces of metal. The total length of the conveyor belt is an astounding 1.2 kilometers. A typical company is said to reduce or recycle between 70 and 80 percent of the waste. Ishizaka's plant has raised this level to more than 95 percent.

6 "We see this plan as something that transforms waste into resources. Recently we were able to process or recycle 97 percent of the waste that came in. We owe such results to that mindset," Ishizaka says. She is now a leading figure in the waste-processing industry, but the road to success was long and bumpy.

7 In 1970, industrial waste processing companies in Japan began converging on the area because of its proximity to Tokyo. Incineration was more common than recycling. At one point in the 1990s, the town had more than 50 smokestacks. In 1999, media reported that vegetables in the area were heavily contaminated with dioxin. The bad news caused vegetable prices to drop sharply. Later, the amount of dioxin detected was not high enough to cause health problems. Nevertheless, the waste processing forms were roundly criticized.

8 Ishizaka's company had already invested more than 13 million dollars in new equipment to prevent dioxin emissions. But since it was the largest in the area,

sort *vt.* 分类；整理
screen *vi.* 筛查；检查
suction *n.* 抽吸；吸力
conveyor belt 传送带
converge *vi.* 聚集，会合
proximity *n.* 邻近
dioxin *n.* 二噁英

its reputation took a hit. Every day, the company faced calls to pack up and leave. At the time, Ishizaka worked in the administration office. She asked her father, who was then CEO, to appoint her to chief operating officer. Later, she succeeded him at the age of 30.

9 "People said they didn't want us there anymore, so that meant we had to find a way of becoming essential to the community," Ishizaka says. "That's when I decided to transform our business into something unlike any other company in the waste processing industry."

10 In a make-or-break challenge, Ishizaka remodeled the entire factory. At a time when the company's annual revenue was around 20 million dollars, she borrowed over 36 million dollars from more than 10 financial institutions and built an environmentally-friendly recycling plant. All of the waste processing was moved indoors to prevent dust and noise pollution. Despite these measures, the residents remained suspicious. Some even speculated the company had moved its operations indoors because it had something to hide.

11 "We were investing large sums of money, so I sometimes wondered why people in the community were still unwilling to accept us. That's when I thought, maybe we should let them see with their own eyes what we were doing here on a daily basis," Ishizaka says. She decided to spend another 1.8 million dollars to create an observation walkway through the facility and she then opened it to the public. Nowadays, the plant welcomes about 10,000 visitors a year, including public officials and entrepreneurs from Japan and abroad.

12 "We're studying the possibility of building a plant with the same capability in Sao Paulo," says Celso Russomanno, a federal congressman from Brazil.

13 "I do lose confidence quite often. We sometimes get complaints and people tell

take a hit 受到损害
remodel *vt.* 重新改造
speculate *vi.* 推测

me my ideas are crazy. But I made a clear choice. So what would I accomplish by quitting now? At this point, I'm determined to finish what I've started," Ishizaka says.

14 The plant occupies 16 hectares of land, 80 percent of which is covered by forest. Today the area is well tended, but it used to look very different. The forest was neglected and littered with illegally dumped waste. Ishizaka rented the land from its owners and set about cleaning it up.

15 "I intentionally didn't buy the land. Instead, I decided to rent it in the hope of fostering a better long-term relationship with the local community. That's how it all began," Ishizaka says. The company even organizes eco-tours through the forest, and the public can enjoy some time off in a green area, with an open square and sports facilities.

16 Yutaka Sekiya owns part of the forest. His family has been farming the area for generations. The tremendous damage caused by the dioxin scandal many years ago led him to play a central role in the campaign to remove Ishizaka's plant from the area.

17 "We even went to court to get the company evicted. But now, the CEO's vision and ethics are just fantastic. I believe our environment is what matters most. If it's harmed in any way, I will rise up again to protect it," Sekiya says.

18 After a new building was completed on the company grounds, ten thousand people, mainly from the local community, visit the site every year. At last, the company once reviled had gained acceptance.

19 "The public needs to understand that the work is actually meaningful for society. I believe our mission is to create and promote an environment that allows people to reuse, reduce and recycle," Ishizaka says.

eco-tour 生态旅游
evict *vt.* 驱逐，逐出
revile *vt.* 辱骂

Unit 6

Waste recycling

In this unit, you will learn:

- **Subject-related knowledge:** The concept of recycling
  Pros and cons of single stream recycling
- **Academic skill:** Documentation: Citing and listing sources
- **Reading strategy:** Making inferences

Section A

Pre-reading

1 Match the words or expressions with the pictures below, and then categorize them into recyclable waste or non-recyclable waste.

medical waste   plastic bottles   kitchen waste
expired medicines   waste paper   waste glass
waste batteries   disposable dishware

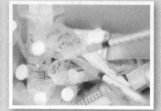

1. _____   2. _____   3. _____

4. _____   5. _____   6. _____

7. _____   8. _____

Recyclable waste: _____
Non-recyclable waste: _____

2 Work in pairs and figure out at least three advantages of waste recycling.

Recycle more? Or... recycle better?

Text A

1 The other day, two colleagues from the waste and recycling world asked me to help settle a dispute. These two very smart people — one with a Ph.D. — were debating the composition of a plastic microwave tray and how it might be recycled... or not. Should they just toss it in the bin for the recycler to deal with? Municipal guidelines were unclear, but it felt wrong to just throw it in the trash. In the end, they tossed it into the single-stream recycling bin and hoped it would be recycled. The episode left me wondering. If even we in the waste management world are so confused, what does this mean for the success of recycling in general?

Changing habits, changing waste

2 Remember newspapers? Once common in our lives, newspapers are increasingly a relic, as more and more of us read our news on computers or portable devices. The result? The United States, for example, generates a whopping 50% less newspaper than they did more than a decade ago, and 20% less paper overall. That's a huge decrease.

3 While paper use has declined, the use of plastics has exploded, with new resins and polymers allowing for new possibilities in packaging. Changing demographics — especially the large Baby Boomer and Millennial generations — mean that more consumers are choosing convenient, single-serve packages for meals and snacks.

4 At the same time, an emphasis on fresh, healthy, and convenient foods is driving a boom in plastic packaging, which can reduce food waste to the tune of preventing 1.7 pounds of food waste for each pound of plastics packaging. However, there is a loss of recyclability on the back end, after those packages have served their use. Today's recycling materials recovery facilities were built to process approximately 80% fiber and 20% containers, not the 40/60 or 50/50 mix. The new mix of inbound material is leading to increased processing costs.

Prevalence of plastic in packaging: Saving grace or problem child?

5 Plastic packaging is both a boon to the environment and a challenge. It's lightweight, great at protecting and preserving goods, and as a petroleum-based product, in the current global marketplace, it's cheap. Flexible plastic packaging — also called "pouches" — offer new levels of convenience and freshness, especially in the food industry.

6 Plastics offer environmental benefits, too. When you look at the entire lifecycle of many types of plastic packaging, they require far fewer raw materials and less energy to manufacture than other packaging alternatives. Over the years, as the use of plastics has grown in consumer goods and packaging — increasingly crowding out glass, metals, and some paper — society has reaped these benefits. Yet, if there is a downside to plastics, it's that they have had a dampening effect on recycling quality.

Plastic Resin Identification Codes

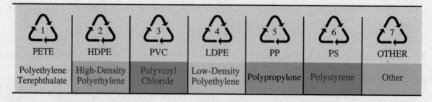

7 It's extremely confusing for consumers to understand how to recycle plastics. For starters, there are all the numbers — 1 through 7 — each with their own, distinct chemical properties, uses, and recyclability. In addition, many

manufacturers are including additives to color their packaging, resulting in low-grade plastics that can't be recycled. So, for every plastic container that can conceivably be recycled — where facilities exist — there are scores that can't.

8 As we all know from experience, it can be mighty challenging indeed to determine how and where to properly recycle these materials. Many of us simply toss plastics into the recycling bin — and hope for the best.

Contamination of the recycling stream

9 The mixing of non-recyclable plastics into the recycling stream — called contamination — is a common occurrence. Types of low-grade plastics are not recyclable, while plastic bags are typically only recyclable by returning them to grocery or retail store for recycling (not curbside), so their presence increases contamination and the cost of recycling across the board. In most communities, an inbound ton of waste now has an average of 17% contamination, while some loads can contain as much as 50% non-recyclable material.

Lightweighting adds to the mix

10 Lightweighting — using lighter material for a product or reducing the weight of the material itself — is becoming a common practice, especially with water bottles made from PET (polyethylene terephthalate). With lightweighting, a typical water bottle now weighs about 37% less than it used to. Lightweighting has many benefits, like reducing the amount of plastics used and therefore produced, and helping to lower freight costs during transport of products. But lightweighting challenges the current economics of recycling.

11 For example, in the current recycling commodities market, recovered PET plastic feedstock is sold by weight, not volume. This means that we need to process 35,000 more bottles than we used to, in order to create one ton of PET feedstock. Using this formula, we would have to process 3.6 billion more water bottles each year to get the same weight of material that we sold a decade ago. Since our costs are currently based on volume and our revenue

based on weight, lightweighting drives up our costs and dampens the long-term economic feasibility of recycling and recovery programs.

What is the end goal?

12 As a society, we used to think that if recycling is good, then more recycling is better. We made recycling convenient so we could collect more recyclables and achieve our weight-based goals. In the process of pushing for higher recycling weights, however, many have lost sight of the actual goal: to lessen the overall environmental impacts of the waste we produce. If we go back to this larger picture, we see that success doesn't necessarily mean recycling large percentages of material based on the weight of the waste stream; rather, success means a reduction in greenhouse gas emissions or raw materials extraction. Recycling is one way to achieve this goal, but it is not the ultimate goal. As we all strive to achieve our overall environmental goals, recycling is just one tool in our toolbox.

13 The popular, newer, non-recyclable plastics test the very goals of recycling. No one wants to put more plastic in a landfill, but when you look at the true environmental impact of different plastic products and uses, the results might surprise you. For example, an EPA lifecycle study looked at different types of coffee packaging to see which consumed the most energy, emitted the most CO_2 equivalent gas, and produced the most municipal solid waste. The researchers found that both the traditional recyclable steel can and the large plastic recyclable container performed worse than the non-recyclable flexible plastic pouch. The lightness and flexibility of the plastic meant such savings in transportation and efficiency that it had a smaller environmental footprint overall than did the recyclable materials.

So, where do we go from here?

14 Flexible plastics aren't going anywhere, so consumer education is crucial to preventing contamination at recycling facilities. This is why recyclers and cities are devoting large amounts of resources to help people understand what's

recyclable — and what's not. We will eventually figure out how to recycle flexible plastics. However, we also need to rethink what our goals really are — and how to best measure them. Is measuring recycling percentages based on weight the best way? Or is it time to find a new way to gauge our success?

15 At waste management, we believe that it is time to change our collective thinking around this critical issue. As the waste stream is increasingly filled with more energy-efficient and lighter weight materials, it's simply not sustainable to continue to set recycling goals that are unrealistic and fail to capture important environmental benefits like overall emissions reductions.

16 Perhaps the time has come to shift to a new metric: a "per capita reduction goal" that can better account for the full value of waste reduction. That's to say, what if the focus weren't just how much you recycle, but how much greenhouse gas you avoid? Instead of recycling for the sake of reaching a weight target, the goal would be to achieve the best overall result for the environment.

17 Such a measure, reflecting a lifecycle approach to managing materials, could go a long way toward accurately capturing the full picture of materials use, and send the right signal for truly sustainable materials management practices.

New words and expressions

dispute /dɪˈspjuːt/ *n.*
a serious argument or disagreement 争论；争辩

tray /treɪ/ *n.* 托盘；浅盘

toss /tɒs/ *vt.*
to throw sth., especially sth. light, without much force 扔，掷，抛（尤指较轻的东西）

episode /ˈepɪsəʊd/ *n.*
an event or a short period of time during which sth. specific happened 一次事件；一段经历

relic /ˈrelɪk/ *n.*
an object from the past that has been kept 遗迹；遗物

whopping /ˈwɒpɪŋ/ *adj.*
very large 巨大的；庞大的

polymer /ˈpɒlɪmə(r)/ *n.* 聚合物；多聚体

demographic /ˌdeməˈɡræfɪk/ *n.*
a group of people in a society, especially people in a particular age group（尤指特定年龄段的）人群

fiber /ˈfaɪbə(r)/ *n.* 纤维；纤维物质

inbound /ˈɪnbaʊnd/ *adj.*
travelling toward a station or an airport 入站的；入境的；归航的

boon /buːn/ *n.*
sth. that is very helpful and makes life easier 恩惠；福利；便利

pouch /paʊtʃ/ *n.* 小袋子；荷包

alternative /ɔːlˈtɜːnətɪv/ *n.*
sth. that you can choose to do or use instead of sth. else 可供选择的事物

reap /riːp/ *v.*
to obtain sth., especially sth. good, as a direct result of sth. that you have done 取得（成果）；收获

downside /ˈdaʊnsaɪd/ *n.*
the negative part or disadvantage of sth.（某物）消极的一面；负面

dampen /ˈdæmpən/ *vt.*
to make sth. weaker or lower in amount 使虚弱；使减少；抑制

additive /ˈædɪtɪv/ *n.*（尤指食品的）添加剂，添加物

conceivably /kənˈsiːvəblɪ/ *adv.*
within the realm of possibility 可以想象地

curbside /ˈkɜːbsaɪd/ *n.* 靠近路缘的人行道部分；路边

freight /freɪt/ *n.*
goods that are transported by ships, planes, trains or lorries / trucks; the system of transporting goods in this way（海运、空运或陆运的）货物；货运

commodity /kəˈmɒdətɪ/ *n.*
a product or a raw material that can be bought and sold 商品

feedstock /ˈfiːdstɒk/ *n.* 原料；给料（指供送入机器或加工厂的原料）

revenue /ˈrevənjuː/ *n.*
the money that a government receives from taxes or that an organization, etc. receives from its business 财政收入；税收收入；收益

feasibility /ˌfiːzəˈbɪlətɪ/ *n.*
the quality of being doable 可行性；可能性；现实性

extraction /ɪkˈstrækʃən/ *n.*
the process of removing an object or substance from sth. else 提炼；取出

ultimate /ˈʌltɪmət/ *adj.*
happening at the end of a long process 最终的

strive /straɪv/ *vi.*
to make a lot of effort to achieve sth. 努力；力求

equivalent /ɪˈkwɪvələnt/ *adj.*
equal in value, amount, meaning, importance, etc.
（价值、数量、意义、重要性等）相等的，相同的
gauge /geɪdʒ/ *vt.*
to make judgement about sth. 判断；判定
metric /ˈmetrɪk/ *n.* 度量标准
crowd out to force sb. or sth. out of a place or situation 把…挤出；排挤

drive up to make prices, costs, etc. increase 使（价格、成本等）上升
single-stream recycling 单流式回收
to the tune of used for emphasizing how large an amount is （数量）高达，多达
energy-efficient *adj.* 能效高的
per capita /pəˈkæpɪtə/ *adj.*
for each person 每人的；人均的

Reading comprehension

Text A can be divided into six parts. Now write down the paragraph number(s) of each part and then fill in the blanks with the missing information.

Part	Paragraph(s)	Main idea
I	Para(s)._____	A debate on _____ between two colleagues aroused the author's reflection.
II	Para(s)._____	With the change of _____, the waste produced is also changed.
III	Para(s)._____	Advantages and disadvantages of _____ _____.
IV	Para(s)._____	_____ may occur in the recycling stream.
V	Para(s)._____	_____ challenges the current economics of recycling.
VI	Para(s)._____	The ultimate goal of recycling is not recycling more but _____.

Unit 6 Waste recycling 115

Language focus

1 Match the words and phrases in Column A with the definitions in Column B and give their Chinese meanings in Column C.

Column A	Column B	Column C
1. energy-efficient	a. state of polluting or being made polluted	_____
2. contamination	b. raw material to supply or fuel a machine or industrial process	_____
3. feasibility	c. of bad quality	_____
4. feedstock	d. the process of removing an object or substance from sth. else	_____
5. low-grade	e. using relatively little energy to provide the power needed	_____
6. alternative	f. the chances that sth. has of happening or being successful	_____
7. extraction	g. sth. that you can choose to do or use instead of sth. else	_____
8. episode	h. an event or a short period of time during which sth. specific happened	_____

2 Fill in the blanks with the words in brackets. Change the form if necessary.

1. The city changed from multi- to single-stream collection a few years ago, now boasts a _____ rate of 69% — one of the highest in the country. (recycle)
2. People are always looking for more convenience — their lives are busy, and they're constantly on the go. Therefore, single-serve _____ foods come in handy. (package)
3. Plastic is a great material for _____ because of its lightweight construction, low cost, and recyclability. (contain)
4. Countries with significant primary industries, such as mining or forestry, tend to _____ far greater quantities of greenhouse gases. (emission)
5. The per capita _____ goal of this city is to remain one of the lowest in the country. (reduce)
6. Some people are suffering ill effects from the _____ of their water. (contaminate)

3 Fill in the blanks with the words given below. Change the form where necessary.

reap   recycler   additive   contamination   landfill
footprint   sustainable   ultimate   freight

1. We all need to look for ways to reduce our carbon _____.
2. The Tyre Recovery Association operates a responsible _____ scheme which aims to ensure that all their members dispose of tyres in a way that is environmentally friendly.
3. _____ is a method of getting rid of waste by burying it in a large deep hole.
4. It is still not possible to predict the _____ outcome of this experiment.
5. We will all _____ the benefits of protecting our environment.
6. By bathing in unclean water, they expose themselves to _____.
7. The country attaches great importance to _____ development of its economy.
8. Researching and developing natural food _____ has already become a trend.
9. The medical supplies were shipped by air _____ to Jordan.

4 Translate the following passage into English.

循环利用可以节约能源，减少原材料的开采，并对抗气候变化。目前，大部分垃圾都是被埋在垃圾填埋场，不仅污染土壤和水源，所释放的气体还会改变气候。所以我们必须大幅减少可生物降解垃圾的填埋数量，同时，加强对垃圾的循环利用，因为这有助于减少人们对各种原材料的需求，从而减少对环境的影响。

Critical thinking

Think about the following questions and give your own answers.

1 Many people say that we have developed into a "throw-away society" because we are filling up our environment with so much rubbish that we cannot fully dispose of. To what extent do you agree with this opinion and what measures can you recommend to help solve this problem?

2 Recently, environmentalists have encouraged consumers to follow the Five R's: Refuse, Reduce, Reuse, Recycle and Rot, which means refusing what you don't need, reducing what you do need, reusing by avoiding disposables and buying second-hand, recycling what you can't reuse, and rotting (composting) what's left. What do you think of this principle? Is it easy to carry it out in your daily life?

Research task

Academic skill: Documentation: Citing and listing sources

I. Introduction

Documentation is the information in the thesis that tells what sources you have used and where you found them.

Documentation has two major tasks: Citing sources and listing sources. For these two tasks, there are numerous styles and ways of giving documentation. In English there is more than one popular format of reference: CMS, MLA, APA, IEEE, FTP, the Vancouver Style, to name but a few. Sometimes a particular format is selected and prescribed for you. Each journal may have its own format and style presented in its style sheet. When you are doing your documentation, make sure what is the required format and style, and then be consistent in your thesis for citing and listing sources.

II. Quotation

Quotations are important in academic writing. When you write an academic paper, you need to borrow ideas, statistics, and previous research findings to strengthen your ideas, explain your arguments, and support your evidence. This practice not only fits your research into the larger picture of a particular field of research, but also lends a basis and persuasiveness to your arguments.

Quotations in a research paper may be either direct or indirect. Both types require documentation. You must provide the sources of both of them.

Direct quotation

A direct quotation is exactly the same as the original and must be put in quotation marks.

1. When to use direct quotations?
    (1) When the original wording expresses the idea so precisely and succinctly that we cannot improve on it;
    (2) When the direct quotation comes from an authority in a particular field and therefore strengthens our point of view;
    (3) When we need to borrow a special term or expression that is a particular writer's invention and carries special meaning;
    (4) When we need the original wording for the purpose of further discussion.

2. How to cite direct quotations?
    (1) Run-in quotation is used when the quotation is fewer than four typed lines or about 40 words. The quotation should run into the text and should be enclosed in double quotation marks. For example:

    Current government policy for England states that "… better collection and treatment of waste from households and other sources has the potential to increase England's stock of valuable resources whilst also contributing to energy policy. And achieving both of these aims helps reduce greenhouse gas emissions."

    (2) Set-off quotation is used when the quotation is more than four lines. Set off a long quotation in an indented block-style paragraph and put the quotation in italics. For example:
    The U.S. Environmental Protection Agency reports that "industrial nonhazardous waste" or "industrial solid waste":

    consists primarily of manufacturing process wastes from sectors such as organic and inorganic chemicals, primary iron and steel, plastics and resin manufacturing, stone, clay, glass and concrete, pulp and paper, and food and kindred products, including wastewater and non-wastewater sludges and solids, and construction and demolition materials.

Indirect quotation

An indirect quotation is to paraphrase or summarize the opinion that you get from the sources in a different and usually simpler way. Academic writers often employ indirect quotations to borrow ideas and previous research findings.

1. The advantage of indirect quotations
    Compared with direct quotations, indirect quotations are more flexible and used more often. This flexibility exists in at least three different ways:

    (1) Flexibility in elaborating the original idea if necessary for better and clearer understanding;
    (2) Flexibility in emphasizing different aspects of the original material to suit a particular purpose;
    (3) Flexibility in modifying the tone and style of the original source for a better fit into the context of the writing.

2. How to cite indirect quotations?
    When you use indirect quotations, the following points should be kept in mind: (1) Don't use quotation marks; (2) Use three spaced periods (…)

to show omissions; (3) Use square bracket ([ ]) for your own words with quotations; (4) Do not change your source; (5) Do not overquote (no more than 30%). For Example:

As a result, a toolbox of cement systems is being developed including geopolymers, CSA cements, and alkali-activated systems with at least one suitable for all waste types[10].
N.B. Milestone, Reactions in cement encapsulated nuclear wastes: need for a toolbox of different cement types, Adv. Appl. Ceram. 105 (1) (2006) 13-20.

III. Citing and listing sources

The process of placing the citation into your text is called "citing a source". The task of citing sources is to indicate the source and related information of each quotation in your thesis. Furthermore, each time you cite a source in your thesis, you need list it in your bibliography. As mentioned in the part of introduction, there are many different formats of citing and listing sources. Here are some examples of the Chicago Manual Style.

The Chicago Manual Style consists of two basic documentation styles: notes and bibliography and author-date. Choosing between the two often depends on the subject matter and the nature of the sources cited, as each system is favored by different groups of scholars.

The notes and bibliography style is preferred by many in the humanities, including those in literature, history, and the arts. This style presents bibliographic information in notes and, often, a bibliography. It accommodates a variety of sources, including esoteric ones less appropriate to the author-date system.

The author-date style has long been used by those in the physical, natural, and social sciences. In this style, sources are briefly cited in the text, usually in parentheses, by author's last name and date of publication. The short citations are amplified in a list of references, where full bibliographic information is provided.

Notes and bibliography style

CMS notes and bibliography style consists of two parts:

1. A superscript number in the text and a corresponding note. (If the note uses a source that has been fully cited previously, shorten the citation using the author's last name and a shortened title.)
2. A bibliography.

The following examples illustrate citations from different sources using the notes and bibliography style.

Source	Bibliography	Footnote
Book	Smith, Zadie. *Swing Time*. New York: Penguin Press, 2016.	**Full Note:** Zadie Smith, *Swing Time* (New York: Penguin Press, 2016), 315–16. **Shortened Note:** Smith, *Swing Time*, 320.
Journal	Satterfield, Susan. "Livy and the Pax Deum." *Classical Philology* 111, no. 2 (April 2016): 165–76.	**Full Note:** Susan Satterfield, "Livy and the *Pax Deum*," *Classical Philology* 111, no. 2 (April 2016): 170. **Shortened Note:** Satterfield, "Livy," 172–73.
Secondary Sources	Gabriel, Astrik L.. "The Educational Ideas of Christine de Pisan." *Culture and Imperialism. Journal of the History of Ideas* 16, no. 1 (1995). Quoted in Sarah Gwyneth Ross. *The Birth of Feminism: Women as Intellect in Renaissance Italy and England*. Cambridge: Harvard University Press, 2009, 23.	**Full Note:** Astrik L. Gabriel, "The Educational Ideas of Christine de Pisan," *Journal of the History of Ideas* 16, no. 1 (1995): 3-21, quoted in Sarah Gwyneth Ross, *The Birth of Feminism: Women as Intellect in Renaissance Italy and England* (Cambridge: Harvard University Press, 2009), 23. **Shortened Note:** Gabriel, "The Educational Ideas," 3-21.
Website	Google. "Privacy Policy." Privacy & Terms. Last modified April 17, 2017. https://www.google.com/policies/privacy/.	**Full Note:** "Privacy Policy," Privacy & Terms, Google, last modified April 17, 2017, https://www.google.com/policies/privacy/. **Shortened Note:** Google, "Privacy Policy."

Author-date style

CMS author-date system consists of two parts:

1. An in-text citation.
2. A reference list.

The in-text citation points the reader to the full information about the source found in the reference list.

The following examples illustrate citations using the author-date system. Each reference list entry is accompanied by a corresponding parenthetical citation.

Source	Reference list	In-text citation
Book	Smith, Zadie. 2016. *Swing Time*. New York: Penguin Press.	(Smith 2016, 315–16)
Journal	LaSalle, Peter. 2017. "Conundrum: A Story about Reading." *New England Review* 38 (1): 95–109. Project MUSE.	(LaSalle 2017, 95)
Secondary Sources	Costello, Bonnie. 1981. *Marianne Moore: Imaginary Possessions*. Cambridge, MA: Harvard University Press.	(quoted in Costello 1981)
Website	Google. 2017. "Privacy Policy." Privacy & Terms. Last modified April 17, 2017. https://www.google.com/policies/privacy/.	(Google 2017)

Task

Search some academic papers on waste recycling and try to find five direct quotations and five indirect quotations in them.

Section B

Reading strategy

Making inferences

The ability to make inferences is, in simple terms, the ability to use several pieces of information from a text in order to arrive at another piece of information that is implicit. Inference can be as simple as associating the pronoun "he" with a previously mentioned male person, and figuring out the meaning of an unfamiliar word as we have discussed in previous units. Or, it can be as complex as understanding a subtle, implicit message, conveyed through the choice of particular vocabulary by the author and drawing on the reader's own background knowledge. Inferring skills are important for reading comprehension, and also more widely applied in the area of literary criticism and other approaches to studying texts.

Proficient readers can often understand when the author tries to tell more than he / she actually say with words. The author gives you hints or clues that enable you to draw conclusions from information that is implied. Using these clues to "read between the lines" and reach a deeper understanding is an important skill for understanding the text, as authors often imply themes and ideas without stating them outright.

Predicting is looking forward while inferring is looking back. Sometimes inferring is harder because readers have to be more precise. Predicting is something one can check his accuracy in further reading, but inferring is not as easy.

Example

What can we infer from the sentences (Para. 1, Text A) "If even we in the waste management world are so confused, what does this mean for the success of recycling in general?"?

Hints

We can infer from the context that the author and his/her colleagues are professionals from the waste and recycling field, but even they don't know how to recycle a plastic microwave tray. That means to correctly recycle the waste is sometimes confusing for common people. That is also the topic the author wants to talk about in this article.

Task

Now read Text B and make out the underlying meanings of statements.

Single stream recycling

Text B

1 Recycling bins around the world have evolved for the 21st century. Instead of diligently separating recyclables into two "streams" – mixed paper (newspaper, junk mail, etc.) and commingled containers (bottles, cans, etc.) – recyclers are now able to put these two streams together in one bin. The new program is called "single stream" recycling. It's the future for responsible resource conservation. With all your recyclables collected in one can, communities and recycling haulers can plan to use the second can for compostable materials like food scraps and yard waste, making it possible for you to recover up to 80% of your discards. Single stream is new. It's different from how we've collected recyclables over the past decades, and there are a lot of questions associated with it.

2 It emerged as an approach that offered greater efficiencies in seemingly every category. Single stream recycling refers to a system in which all kinds of recyclables such as plastics, paper, metals, glass etc. are put into a single bin by consumers. Afterward, the recyclables are collected and transported to a Material Recovery Facility (MRF) where they are sorted and processed. A benefit of this approach is that consumers or the depositors of the commodities don't have to separate or sort the recyclables. Rather, they are encouraged to put everything that is not trash into a single bin. This approach helps to increase the quantity of material recovered, but, as we will discuss below, not necessarily the quality.

commingled *adj.* 混合的
hauler *n.* 运输者
compostable *adj.* 可堆肥的

History of single stream secycling

3 In the 1990s, several California communities started using single stream recycling and subsequently the system was adopted by communities across the United States. By 2005, around one-fifth of all locations with recycling programs in the U.S. were employing the single stream recycling system. By the start of the current decade, the number reached more than two-thirds. By 2012, around 248 MRFs in the U.S. used a single stream recycling system.

How it works

4 Once the recyclables are put into curbside recycling bins, MRFs collect, sort, and process the recyclables. After processing, similar kinds of recyclables are baled and shipped to recyclers of specific materials, ultimately to be utilized in the production of new products. This is a very simple description of the process.

5 The actual sorting process may vary with respect to the automation employed in the system, involving such technologies as conveyors, screens, forced air, magnets, optical material identification and eddy current.

6 Let's take a look at one example. First of all, all the materials are unloaded and placed on a conveyor. Initially, non-recyclable items are manually sorted and removed. After the initial sorting is completed, the materials move to a triple-deck screen. There, all cardboard, containers and paper — items too heavy or too light for the next level of single stream recycling process, are removed. Heavier containers drop to the bottom level while lighter items head to the second. This screen also breaks the glass containers for the safety and convenience of the workers.

bale vt. 将…捆成大包
automation n. 自动化
eddy current 涡电流
manually adv. 手动地

7 The remaining materials pass under a powerful magnet to remove tin and steel cans. Next, MRF staff watch carefully for specific commodities that may still have inadvertently made it down the line. Lastly, a reverse magnet called eddy current causes the aluminum cans to fly off the conveyor and into a bin. Various types of fiber are separated in the single stream MRF. MRF workers separate cardboard, newsprint, office paper and drop each piece into a bunker below. Once all the materials are separated, the materials are baled and shipped to recycling companies for processing.

8 The whole process of single stream recycling involves a combination of machine and human workers. <u>The industry trend is towards state-of-the-art MRFs and a move away from legacy or "dirty" MRFs which are much more labor-intensive.</u>

Advantages of single stream recycling

9 One of the most notable benefits of single stream recycling is increased recycling rates. As the individuals or consumers don't have to do the sorting, they are more encouraged to participate in the curbside recycling programs. Again, less space is required to store collection containers. From the point of view of the collection, costs for the hauling process are reduced versus separate pickups for different recycling streams, or the hauler having to place different materials into various truck compartments. The simple process receives greater public approval.

Disadvantages of single stream recycling

10 The most notable criticism of single stream recycling system is that it led to a decrease in the quality of materials recovered. Putting all the materials into a single bin surely increases the level of contamination in terms of such problems

---

inadvertently *adv.* 不经意地，疏忽地
bunker *n.* 料仓；贮仓

as broken glass and the propensity to toss non-approved materials into the recycling bin, ultimately causing significant problems for MRF operators and communities. Although the consumers or depositors are not sorting the materials, someone ultimately has to sort them, making the cost of recycling higher. Ultimately, the public convenience comes at a cost.

11 Such an approach had shortcomings. These included:
- Because of compartment limitations on collection trucks, the opportunity for the introduction of new scrap materials for recycling is limited.
- Because materials had to be dumped into the appropriate compartment of the truck, there was more pedestrian travel time and handling at each stop.
- The extra handling at each stop not only translated into more labor, as addressed above, but also less efficient utilization of trucks.
- Another source of truck inefficiency was that if one compartment filled up, the truck would have to dump it before continuing.
- More bins for segregation translated into more complexity and space requirements for recycling participants, and the greater likelihood of more recyclable material just ending up in the trash.

12 There are both advantages and disadvantages of single stream recycling — and the battle is between quality and convenience. To this point in time, the popularity of single stream suggests that convenience has trumped quality. The use of single stream has become increasingly controversial, however, in the wake of more strict quality requirements for recycled materials. At least one community has reverted to a dual stream collection system to help it reduce costs.

propensity *n.* 倾向，习性
trump *vt.* 战胜，打败
controversial *adj.* 有争议的
revert *vi.* 恢复，还原

Unit 7

Ecosystems

In this unit, you will learn:

- **Subject-related knowledge:** Types of ecosystems and knowledge about ecology
- **Academic skill:** Making a presentation (Part I)
- **Reading strategy:** Making prediction

Section A

Pre-reading

1 Match the following expressions with the pictures below.

freshwater ecosystem   grassland   tundras
marine ecosystem   savannas   amphibian

1. _____   2. _____   3. _____

4. _____   5. _____   6. _____

2 Discuss the following questions with your partner.

1. What types of ecosystems have you seen?
2. How can we protect ecosystems?

Major types of ecosystems

1 An ecosystem is a community of plants and animals interacting with each other in a given area, and also with their nonliving environments. Ecosystems can be huge, with many hundreds of different animals and plants all living in a delicate balance, or they could be relatively small. In particularly harsh places in the world, particularly the North and South Poles, the ecosystems are relatively simple because there are only a few types of creatures that can withstand the freezing temperatures and harsh living conditions. Some creatures can be found in multiple different ecosystems all over the world in different relationships with other or similar creatures. Ecosystems also consist of creatures that mutually benefit from each other. For instance, a popular example is that of the clown fish and the anemone — the clown fish cleans the anemone and keeps it safe from parasites as the anemone stings bigger predators that would otherwise eat clown fish.

2. There are very many types of ecosystems out there, but the three major classes of ecosystems, sometimes referred to as "biomes", which are relatively contained, are freshwater ecosystems, terrestrial ecosystems and marine ecosystems.

Freshwater ecosystems

3. As is clear from the title, freshwater ecosystems are those that are contained to freshwater environments. This includes, but is not limited to, ponds, rivers and other waterways that are not the sea. Freshwater ecosystems are actually the smallest of the three major classes of ecosystems. The ecosystems of freshwater systems include relatively small fish (bigger fish are usually found in the sea), amphibians (such as frogs, toads and newts), insects of various sorts and, of course, plants. Freshwater ecosystems can also include birds, because birds often hunt in and around water for small fish or insects. The absolutely smallest living part of the food web of these sorts of ecosystems is plankton, a small organism that is often eaten by fish and other small creatures.

4. Freshwater ecosystem can be broken up into smaller ecosystems. For instance, there are:
   - **Pond ecosystems** — These are usually relatively small and contained. Most of the time they include various types of plants, amphibians and insects. Sometimes they include fish, and most of the time fish are artificially introduced to these environments by humans.
   - **River ecosystems** — Because rivers always link to the sea, they are more likely to contain fish alongside the usual plants, amphibians and insects.

Terrestrial ecosystems

5. Terrestrial ecosystems are varied because there are so many different sorts of places on Earth. Some of the most common terrestrial ecosystems are:
   - **Tundras** — Tundras are generally found north of the boreal forests. These regions are known for the cold as they have mostly frozen subsurface soil and permafrost. They lack trees and vegetation grows short and only

when enough of the topsoil has thawed. Tundras have a similar level of precipitation as deserts. Organisms that live in this ecosystem are highly adapted to the harsh cold. They include things like reindeers, snow owls, and geese. You will not find any reptiles in this region. Tundras found in mountainous regions are generally referred to as alpine tundras. Organisms that live here are generally migratory as they move to the alpine tundras during warmer times and leave when it gets cold.

- **Deserts** — Deserts are ecosystems with hardy inhabitants, able to survive in an environment that receives less than 10 inches of rainfall annually. Deserts may be hot or cold. The desert is home to many plants that lie dormant until it rains, when they bloom and spread their seeds, then lie dormant until the next major rainfall. Deserts are also home to plants capable of storing their own water, such as cacti. Other plant adaptations in deserts include widespread roots and small leaves with waxy coverings. In hot deserts, some desert animals survive the searing heat by burrowing or living in caves. Many animals are largely nocturnal, staying underground during the heat of the day and foraging for food at night when it is cooler.
- **Forests** — There are many different types of forests all over the world, including deciduous forests and coniferous forests. Forests can support a lot of life and can have very complex ecosystems. Forests can be divided further into four different subgroups, but they all terrestrial ecosystems have a dense tree population and medium to high levels of precipitation in common. Tropical rain forests are home to a great diversity of animals. The climate is hot with excessive rainfall, and vegetation grows in several layers from the forest floor to the canopy. The forests of India and eastern Brazil, however, have specific seasons of rain and dry weather. These forests are called tropical deciduous forests. Coastal coniferous and temperate deciduous forests flank the west and east coasts of the U.S., respectively. They experience four seasons, and only moderate rainfall. Temperate rain forests also occur along the northwest coast of North America. The northern Canadian forests are predominantely coniferous and experience long sub-arctic winters.

- **Grasslands** — In a grassland ecosystem, trees are scarce, removed by environmental conditions and brush fires. However, the grasslands, as their name indicates, receive sufficient precipitation to sustain different varieties of grasses. Today, many grasslands are becoming endangered because of farming practices and grazing herds of animals, especially when overgrazing occurs. The grasslands are subdivided into tropical grasslands (also known as the savannas); temperate grasslands, like the prairies of the Midwest in the U.S.; and the polar grasslands like the northern Canadian tundra. Savannas generally receive 20 to 50 inches of rain per year, concentrated in a six- to eight-month span, followed by a dry season. Temperate grasslands have hot summers and cold winters, with average annual rainfalls between 20 and 35 inches.

6 Because terrestrial ecosystems are so diverse, it is difficult to make generalizations about them. However, a few things are almost always true. For instance, most of them contain herbivores that eat plants and all have carnivores that eat herbivores and other carnivores. Some places, such as the poles, contain mainly carnivores because no plant grows. A lot of animals and plants that grow and live in terrestrial ecosystems also interact with freshwater and sometimes even ocean ecosystems.

Marine ecosystems

7 Marine ecosystems are relatively contained, although they, like freshwater ecosystems, also include certain birds that hunt for fish and insects close to the ocean's surface. Marine ecosystems have a high level of salt and comprise the many oceans of the world and other bodies of saltwater. Because of the biodiversity and size of marine ecosystems, they are the most abundant ecosystems found in the world.

8 The pelagic marine ecosystem represents the open ocean where marine life swims freely or float. They are not attached to the bottom or any surfaces. These can be things like plankton or whales. Benthic marine ecosystems represent the bottom of the ocean where organisms are attached to something

or very close to the bottom. These can include corals or mangroves regions, each of which has an ecosystem of their own. These two sub-ecosystems have further subsequent ecosystems that are defined by many of the listed factors above.

9 There are different sorts of marine ecosystems:
   - **Shallow water** — Some tiny fish and coral only live in the shallow waters close to land.
   - **Deep water** — Big and even gigantic creatures can live deep in the waters of the oceans. Some of the strangest creatures in the world live right at the bottom of the sea.
   - **Warm water** — Warmer waters, such as those of the Pacific Ocean, contain some of the most impressive and intricate ecosystems in the world.
   - **Cold water** — Less diverse, cold waters still support relatively complex ecosystems. Plankton usually forms the base of the food chain, followed by small fish that are either eaten by bigger fish or by other creatures such as seals or penguins.

10 Marine ecosystems are amongst some of the most interesting in the world, especially in warm waters such as those of the Pacific Ocean. This is because around 75% of the Earth is covered by the sea, which means that there is lots of space for all sorts of different creatures to live and thrive. In the oceanic ecosystems the very base of the food chain is plankton, just as it is in freshwater ecosystems.

New words and expressions

anemone /əˈnemənɪ/ n.
(=sea anemone) 海葵

parasite /ˈpærəsaɪt/ n.
a small animal or plant that lives on or inside another animal or plant and gets its food from it 寄生物；寄生虫；寄生植物

amphibian /æmˈfɪbɪən/ n. 两栖动物

plankton /ˈplæŋktən/ n.
the very small forms of plant and animal life that live in water 浮游生物

tundra /ˈtʌndrə/ n. 冻原，苔原（树木不生，底土常年冰冻的北极地区）

permafrost /ˈpɜːməfrɒst/ n. 永久冻土

thaw /θɔː/ vi.
if ice or snow thaws or is thawed, it becomes warmer and turns into water （冰或雪）融化

precipitation /prɪˌsɪpɪˈteɪʃən/ n. （雨、雪等的）降落；降水量

reptile /ˈreptaɪl/ n. 爬行动物

alpine /ˈælpaɪn/ adj.
existing in or connected with high mountains, especially the Alps in Central Europe 高山的；高山上的（尤指阿尔卑斯山的）

dormant /ˈdɔːmənt/ adj.
not active or not growing right now, but able to be active later 蛰伏的；休眠的

cactus /ˈkæktəs/ n. (pl. cacti, cactuses) 仙人掌

waxy /ˈwæksɪ/ adj.
smooth and shiny like wax 像蜡一样光滑的

searing /ˈsɪərɪŋ/ adj.
extremely hot 炽热的；灼热的

nocturnal /nɒkˈtɜːnəl/ adj.
an animal that is nocturnal is active at night （动物）夜间活动的，夜行的

deciduous /dɪˈsɪdʒʊəs/ adj. （树）落叶的

coniferous /kəʊˈnɪfərəs/ adj. （树）针叶的

canopy /ˈkænəpɪ/ n. 树冠

flank /flæŋk/ vt.
to be at the side of sth. or sb. 位于…的侧面

predominantly /prɪˈdɒmɪnəntlɪ/ adv.
mainly or mostly 主要地；绝大多数地；大部分地

graze /greɪz/ vt.
to put an animal in a place where it can eat grass 放牧；放…啃食牧草

savanna /səˈvænə/ n.
a large flat area of grassy land in a warm part of the world（热带或亚热带的）稀树草原

prairie /ˈpreərɪ/ n.
a wide open area of mostly flat land that is covered with grass, especially in North America （尤指北美洲的）大草原

herbivore /ˈhɜːbɪvɔː(r)/ n.
any animal that eats only plants 食草动物

sustenance /ˈsʌstɪnəns/ n.
food and drink 食物；营养

carnivore /ˈkɑːnɪvɔː(r)/ n.
an animal (such as a dog, fox, crocodile, or shark) that feeds primarily or exclusively on flesh 食肉动物

pelagic /peˈlædʒɪk/ adj.
relating to or living in the ocean, far from shore 远洋的；深海的

benthic /ˈbenθɪk/ adj.
of, relating to, or occurring at the bottom of a body of water 水底的

Reading comprehension

1 Read Text A and complete its outline.

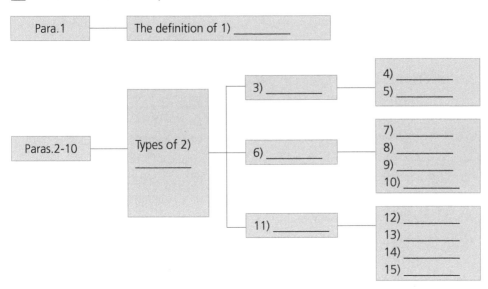

Language focus

1 Match the words and phrases in Column A with the definitions in Column B and give their Chinese meanings in Column C.

Column A	Column B	Column C
___ 1. insect	a. a small creature that has multiple legs and usually have wings	_____
___ 2. herbivore	b. any animal that eats only plants	_____
___ 3. parasite	c. an animal that feeds primarily or exclusively on flesh	_____
___ 4. organism	d. any animal that can live both on land and in water	_____
___ 5. polar	e. connected with, or near the North or South Pole	_____
___ 6. carnivore	f. an organism living in, with, or on another organism	_____
___ 7. amphibian	g. a living thing, especially one that is extremely small	_____
___ 8. prairie	h. wide open area of mostly flat land that is covered with grass, especially in North America	_____

Unit 7 Ecosystems 137

2 Fill in the blanks with the words given below. Change the form where necessary.

terrestrial   ecosystem   precipitation   dormant
polar   grassland   rainforest   temperature

1. To make things worse, people let their cattle eat up the grass, which further destroys the _____.
2. Recently, news stories about _____ bears desperately reaching for an ice floe have become increasingly common.
3. A thin or thick snow cover could change the physical, chemical, and biological properties of the soil by altering the soil _____ and soil moisture, which may influence the cycle of soil carbon.
4. During the winter, the seeds lie _____ in the soil.
5. The death of baobab trees could have far-reaching implications for the _____ in Africa.
6. As _____ beings, it is easy for us to underestimate our dependence on the global ocean.
7. High _____ and cold air combine to create huge glaciers that spill into numerous fjords (峡湾).
8. We were now leaving the dry forest and plunging into the far more biologically diverse _____.

3 Read the text and tell whether the following statements are true or false. If false, underline the mistakes and put the corrections in the blanks provided.

1. Birds are not included in freshwater system. _____
2. Reindeers, snow owls, and reptiles can be found in the tundras. _____
3. Marine eco-system has only two sub-ecosystems. _____
4. In pond ecosystems, fish are mostly natural inhabitants. _____
5. Forests have the highest levels of precipitation among all ecosystems. _____
6. In recent years, many ecosystems, especially grasslands, are becoming destroyed because of human activities. _____
7. Some creatures can be found in multiple different ecosystems all over the world. _____
8. Freshwater ecosystems are actually the biggest of the three major classes of ecosystems, while marine ecosystems the largest. _____

4 Translate the following paragraph into English.

地球上的生态系统种类繁多。生态系统的不同取决于其温度、位置、以及构成等很多因素。我们把生态系统当作研究和理解生物多样性、地域多样性以及如何充分保护这些区域防止其进一步恶化的工具。生态系统是由某个特定区域中存在的生物和非生物定义的。这些区域之间并非总是界线分明,有时随着生态系统中某一部分的移动,这些区域会连接在一起,模糊不清。这些生态系统有森林生态系统、沙漠生态系统、草原生态系统、苔原生态系统、海洋生态系统和淡水生态系统。许多生态系统还可以进一步划分为子系统。河口可以被称为一个生态系统,但是因为它是海洋生态系统与淡水生态系统的混合系统,因此这个问题还存有争议。

Critical thinking

1 Bats, masked palm civets and pangolins are integral parts of nature. They serve well in the balance of the ecosystem. However, some people illegally hunt these wild animals, trade them and make them into food. What risks do you think such practices pose to the ecosystem?

2 Some wild animals are the hosts of viruses. Human beings are susceptible to these viruses and may die of them. What role do you think wild animals play in the balance-keeping of ecosystem? How should human beings get on with nature?

Research task

Academic skill: Making a presentation (Part I)

1. What is a presentation

   A presentation is the process of presenting a topic to an audience. Presentations provide you with an opportunity to show something that you have researched to other people. Usually it is given with the help of PowerPoint slides.

   Some presentations are given in a classroom as the outcome of an assignment a certain area. Other presentations are for more professional environments, like academic or conference presentations.

2. How to prepare a presentation

   The key to making an effective presentation is to be well prepared. Here are a few tips that will make the process smoother for you:

   Structure your presentation well

   Remember that your presentation is meant to be heard, not read. Audiences typically have lower attention spans than readers; therefore, keep the content simple and straightforward. Structure your presentation well, with a clear introduction, body, and conclusion. Open the presentation with greetings or some general remarks such as "Good morning ladies and gentlemen, today I will introduce the project I have done in the following three aspects", etc. In the body part, try to divide your content in several different parts and mention the beginning of each part to remind the audience of the split. Use structural words like "firstly", "secondly", "lastly"; or "in the first part", "next I will say something about my experience", "in the end let's have a look at", etc. And for the conclusion try to provide a quick recap of the main points.

Pay attention to the language

The language on the PowerPoint slides should be very short and concise, preferably like an outline; at the same time, the content should be delivered all by your speech. So use oral language that is simple and clear. Avoid jargons that might confuse your audience.

Make full use of visual aids

Oral presentations can be highly enhanced in effectiveness if you employ some visual aids. The audience can understand your content more easily. But, make sure the visual aids are simple, easy to understand and suitable in color and style; otherwise they will distract your audience from your speech.

Adhere to time limits

Generally, the presentation session is given a certain amount of time, so prepare your material accordingly. Also, be prepared for any last-minute changes in session timings, which means you should be very familiar with your speech and PowerPoint slides so you can make adjustments quickly.

Rehearse your presentation

Rehearsing a few times in front of a friend or in front of the mirror can make you familiar with the content, boost your confidence as well, and help you attain a better control of time.

Task

Please make a 10-minute presentation based on Text A. Remember to follow the above-mentioned five tips. Usually, a presentation consists of three parts: introduction, body and conclusion. In the introduction, you can state your purpose and try to capture the audience's attention. In the body part, you present your main points in a logical order. You may use examples or anecdotes to illustrate your points. In the conclusion, you may summarize the main points you have covered and restate your purpose. You can also give the audience a chance to ask questions.

Section B

Reading strategy

Making prediction

Making predictions is a strategy in which readers use information from a text (including titles, headings, pictures, diagrams, etc.) or their own personal experiences to anticipate what they are about to read or what comes next. Readers involved in making predictions focus on the text at hand, constantly thinking ahead and also refining, revising, and verifying their predictions. This strategy also helps readers make connections between their prior knowledge and the text.

Actually, people make predictions all the time, though they may not realize it. They predict what their friends will say when they score a winning point. They predict what's for dinner when they come home to warm smells from the kitchen. People are able to make predictions based on their prior knowledge, or information they already have. The predicting strategy can also be applied in reading practice.

Before you read a text in detail, it is possible to predict what information you may find in it. You will probably have some knowledge of the subject already, and you can use this knowledge to help you anticipate what the reading text contains. After looking at the title, for example, you can ask yourself what you know and do not know about the subject before you read the text. Or you can formulate questions that you would like otherwise to have answered only by reading the text. These exercises will help you focus more effectively on the ideas in a text when you actually start reading.

Task

Look at the title of Text B and write down your answers to the following questions.

What is the topic?

What do you already know about it?

What else do you want to know about it?

Proficient readers make predictions smoothly, and confirm or disconfirm them based on subsequent reading.

Now please read Text B and try to identify the possible information and framework for the following titles by using the reading skills mentioned above and compare what you predict with the information and framework of Text B.

Title	What is ecology?
Information	
Framework	

Text B

What is ecology?

1. Ecology refers to the set of relationships between organisms and their environment. All ecology breaks down and falls under three basic types. Under each of the basic types you can break down the ecologies even further, but these are the ones that will give you an understanding of what is an ecology.

2. **Conservation ecology** — This refers to the ecology of the natural world and how it exists without the presence or interference of man, and how it exists with the presence and interference of man. It is common to find different areas in the wilderness or portions of public parks set aside with entrance restricted or forbidden because they are conservation areas. Conservation areas are set aside to protect conservation ecologies.

3. **Urban ecology** — As you can guess by the name this one is focused on the balance of life within urban settings. It includes both conservation and human ecology, but the nature of an urban ecology is so different in how phases of ecological succession may develop that it is studied as its own entity. Urban

---

conservation *n.* 保护区；养护

ecology looks at the impact of human life on an area, how cities and urban areas manage resources and what the cycle of growth and decay within these unique environments are.

4 **Human ecology** — Human ecology is multi-faceted. It focuses greatly on patterns of population and mortality, consumption of resources, conservation efforts and how humans affect plant, animal and other human life. Within human ecology you will also find people studying the impact of the human race on the atmosphere, space and the ocean.

Three phases of ecological succession

5 Each of these ecology types go through three distinct phases of ecological succession. Succession is the term used to define how the development in one phase of the ecology is then used to allow the next phase to occur. While it may seem as if this is a very ordered process, if the ecology is disrupted it may go back or forward a phase. To better explain succession we will use the example of what happens after a forest fire has destroyed life on the side of a mountain.

6 • **Primary** — This is the beginning stage of an ecology. That side of the mountain may begin to be repopulated with seeds that have blown in on the wind, been planted by humans in a conservation effort, and the area may gain animal life from different types moving in to hunt or live for protection. The key to understanding primary ecology is that it begins with the minimal life that the area can support, which is then followed by the next life in the food or resource chain.

7 • **Secondary** — During the secondary phase of the ecology life has begun again in the forest area. Grass and flowers may grow, small birds and mammals have returned and there have been two or three cycles of life that have occurred. This then brings in more predators in the cycle to complete the cycle of life.

---

disrupt *vt.* 扰乱；使…混乱
repopulate *vt.* 使重新住入

8. • **Climax** — The climax stage of ecological succession sees a forest thriving on the mountainside again with all the plant and animal life you would expect. There is little evidence of the fire as the forest has aged enough to still be young, but to be well established. It is at this point that the ecology begins to slowly decline as it begins to overuse resources and conservation is needed.

What can disrupt an ecology?

9. Problems happen when an ecology is disrupted. This can speed an ecology through its successive phases and cause it to die out quickly. Remember that no ecology exists by itself, but all ecologies exist within balance to each other. A small conservation ecology in the rainforest may not seem like a big deal, but it plays a part in managing air, climate and other resources needed by the human ecology a continent away.

Natural disruptions

10. Natural disruptions can come in the form of extreme weather such as prolonged droughts, heavy rain or snowfall, storms, hurricanes, tornadoes, monsoons and more. These also can involve seismic disruptions such as earthquakes and volcanic eruptions. These dramatic events can damage or change the balance within an ecology.

Man-made disruptions

11. Man-made disruptions can come in all shapes and sizes. There are the obvious ones such as clearing land, but less obvious ones when you are looking at an ecology in its secondary stage of succession that includes mankind. An increased drain on natural resources, such as water or fossil fuels, can set up a disruption. All forms of mining, including wind farms, disrupt ecologies as well.

---

prolong *vt.* 延长；拖延
monsoon *n.* 季风季节；雨季
seismic *adj.* 地震的；地震性的；地震引起的

How to protect ecology?

12 There are many different ways that an ecology can be protected. Most of them are in place or in the process of being put in place now as the world begins to understand the importance of managing ecologies better. The three main ways that an ecology is protected are:

13 **Conservation** — As mentioned before, conservation is when an ecology is protected from harm by being set aside so it cannot be interfered with. In the U.S., for example, many national parks have established conservation areas. You can only visit certain places in the park while other areas are off limits to protect the ecology there.

14 **Regulation** — Regulations such as the Clean Air Act and other laws try to set limits on the types of man-made disruptions caused by expansion, exploration or industrialization that can occur. The goal is to limit the known impact on the natural ecology to preserve and maintain the resources that mankind depends on.

15 **Replacement** — This type of replacement is twofold. In areas where regulation has permitted conservation to be removed, companies may pay to have a new area seeded or protected to try and restore balance. Another aspect of replacement is the search for man-made substances that can replace natural resources. One example of this is the search for a viable alternative energy to fossil fuel.

---

viable *adj.* 切实可行的；可实施的

Unit 8

Environmental protection

In this unit, you will learn:

- **Subject-related knowledge:** Principles and approaches of environmental protection
  Causes and effects of environmental degradation
- **Academic skill:** Making a presentation (Part II)
- **Reading strategy:** Outlining (Part II)

Section A

Pre-reading

1 Match the following expressions with the pictures listed below.

water pollution   sound pollution   debris flow
quake lake   electromagnetic radiation pollution   air pollution

1. _____   2. _____   3. _____

4. _____   5. _____   6. _____

2 Discuss the following question with your partner.

What can you do to protect environment?

Text A

Principles and approaches of environmental protection

Principles of environmental law

1   Environmental protection is a practice of protecting the natural environment on individual, organization controlled or governmental levels, for the benefit of both the environment and humans. Due to the pressures of overconsumption, population and technology, the biophysical environment is being degraded, sometimes permanently. This has been recognized, and governments have begun placing restraints on activities that cause environmental degradation. Since the 1960s, activity of environmental movements has created awareness of the various environmental problems. There is no agreement on the extent of the environmental impact of human activity and even scientific dishonesty occurs, so protection measures are occasionally debated.

2   Environmental law has developed in response to emerging awareness of and concern over issues impacting the entire world. While laws have developed piecemeal and for a variety of reasons, some effort has gone into identifying

key concepts and guiding principles common to environmental law as a whole. The principles discussed below are not an exhaustive list and are not universally recognized or accepted. Nonetheless, they represent important principles for the understanding of environmental law around the world.

Sustainable development

3 Defined by the United Nations Environment Programme (UNEP) as "development that meets the needs of the present without compromising the ability of future generations to meet their own needs", sustainable development may be considered together with the concepts of "integration" (development cannot be considered in isolation from sustainability) and "interdependence" (social and economic development, and environmental protection, are interdependent). Laws mandating environmental impact assessment and requiring or encouraging development to minimize environmental impacts may be assessed against this principle.

4 The modern concept of sustainable development was a topic of discussion at the 1972 United Nations Conference on the Human Environment (Stockholm Conference), and the driving force behind the 1983 World Commission on Environment and Development (WCED, or Bruntland Commission). In 1992, the first UN Earth Summit resulted in the Rio Declaration, Principle 3 of which reads: "The right to development must be fulfilled so as to equitably meet developmental and environmental needs of present and future generations". Sustainable development has been a core concept of international environmental discussion ever since, including at the World Summit on Sustainable Development (Earth Summit 2002), and the United Nations Conference on Sustainable Development (Earth Summit 2012, or Rio+20).

---

the United Nations Environment Programme 联合国环境规划署
the Rio Declaration《里约宣言》(《里约环境与发展宣言》的简称)

Equity

5 Defined by UNEP to include intergenerational equity — "the right of future generations to enjoy a fair level of the common patrimony" — and intragenerational equity — "the right of all people within the current generation to fair access to the current generation's entitlement to the Earth's natural resources" — environmental equity considers the present generation under an obligation to account for long-term impacts of activities, and to act to sustain the global environment and resource base for future generations. Pollution control and resource management laws may be assessed against this principle.

Transboundary responsibility

6 Defined in the international law context as an obligation to protect one's own environment, and to prevent damage to neighboring environments, UNEP considers transboundary responsibility at the international level as a potential limitation on the rights of the sovereign state. Laws that act to limit externalities imposed upon human health and the environment may be assessed against this principle.

Public participation and transparency

7 Identified as essential conditions for "accountable governments, ... industrial concerns," and organizations generally, public participation and transparency are presented by UNEP as requiring "effective protection of the human right to hold and express opinions and to seek, receive and impart ideas, ... a right of access to appropriate, comprehensible and timely information held by governments and industrial concerns on economic and social policies regarding the sustainable use of natural resources and the protection of the environment, without imposing undue financial burdens upon the applicants and with adequate protection of privacy and business confidentiality", and "effective judicial and administrative proceedings." These principles are present in environmental impact assessment, laws requiring publication and access to relevant environmental data, and administrative procedure.

Precautionary principle

8 One of the most commonly encountered and controversial principles of environmental law, the Rio Declaration formulated the precautionary principle as follows:

9 In order to protect the environment, the precautionary approach shall be widely applied by States according to their capabilities. Where there are threats of serious or irreversible damage, lack of full scientific certainty shall not be used as a reason for postponing cost-effective measures to prevent environmental degradation.

10 The principle may play a role in any debate over the need for environmental regulation.

Prevention

11 The concept of prevention can perhaps better be considered an overarching aim that gives rise to a multitude of legal mechanisms, including prior assessment of environmental harm, licensing or authorization that set out the conditions for operation and the consequences for violation of the conditions, as well as the adoption of strategies and policies. Emission limits and other product or process standards, the use of best available techniques and similar techniques can all be seen as applications of the concept of prevention.

Polluter-pays principle

12 The polluter pays principle stands for the idea that "the environmental costs of economic activities, including the cost of preventing potential harm, should be internalized rather than imposed upon society at large." Many issues related to responsibility for cost for environmental remediation and compliance with pollution control regulations involve this principle.

Approaches with regards to environmental protection
Voluntary environmental agreements

13 In industrial countries, voluntary environmental agreements often provide a

platform for companies to be recognized for moving beyond the minimum regulatory standards and thus support the development of environmental practices. In some countries, these agreements are more commonly used to remedy significant levels of non-compliance with mandatory regulation. The challenges that exist with these agreements lie in establishing baseline data, targets, monitoring and reporting. Due to the difficulties inherent in evaluating effectiveness, their use is often questioned and, indeed, the whole environment may well be adversely affected as a result. The key advantage of their use in these countries is that it helps to build environmental management capacity.

Ecosystems approach

14 An ecosystems approach to resource management and environmental protection aims to consider the complex interrelationships of an entire ecosystem in decision-making rather than simply responding to specific issues and challenges. Ideally the decision-making processes under such an approach would be a collaborative approach to planning and decision-making that involves a broad range of stakeholders across all relevant governmental departments, as well as representatives of industry, environmental groups and community. This approach ideally supports a better exchange of information, development of conflict-resolution strategies and improved regional conservation.

International environmental agreements

15 Many of the Earth's resources are especially vulnerable because they are influenced by human impacts across many countries. As a result of this, many attempts are made by countries to develop agreements that are signed by multiple governments to prevent damage or manage the impacts of human activity on natural resources. This can include agreements that impact factors such as climate, oceans, rivers and air pollution. These international environmental agreements are sometimes legally binding documents that have legal implications when they are not followed and, at other times, are

more agreements in principle or are for use as codes of conduct. These agreements have a long history with some multinational agreements being in place from as early as 1910 in Europe, America and Africa. Some of the most well-known international agreements include the Kyoto Protocol and others.

the Kyoto Protocol《京都议定书》

New words and expressions

piecemeal /ˈpiːsmiːl/ *adv.*
made or done in separate stages rather than being planned and done as a whole 一个一个地；逐一地

integration /ˌɪntɪˈɡreɪʃən/ *n.*
the process of combining with other things in a single larger unit or system 结合；融合；整合

interdependence /ˌɪntədɪˈpendəns/ *n.*
the condition of a group of people or things that all depend on each other 互相依赖

mandate /ˈmændeɪt/ *vt.*
to give an official command that sth. must be done 命令；指示

assessment /əˈsesmənt/ *n.*
an opinion or a judgment about sb. / sth. that has been thought about very carefully 评估；看法

patrimony /ˈpætrɪməʊni/ *n.* 祖传的财产；遗产

entitlement /ɪnˈtaɪtlmənt/ *n.*
the right to receive sth. or do sth.（接受某物或做某事的）权利，资格

impose /ɪmˈpəʊz/ *vt.*
to introduce sth. such as a new law or new system, and force people to accept it 实施，推行（新的法律、制度等）

transparency /trænsˈpærənsi/ *n.*
an honest way of doing things that allows other people to know exactly what you are doing 透明度；公开度

accountable /əˈkaʊntəbl/ *adj.*
responsible for the effects of your actions and willing to explain or be criticized for them 负有责任的

undue /ˌʌnˈdjuː/ *adj.*
more than is reasonable, appropriate, or necessary 不适当的；过度的，过分的

confidentiality /ˈkɒnfɪˌdenʃiˈæləti/ *adj.*
a situation in which you trust sb. not to tell secret or private information to anyone else 机密，秘密；保密

judicial /dʒuːˈdɪʃəl/ *adj.*
connected with a court, a judge or legal judgment 法庭的；法官的；审判的；司法的

proceeding /prəʊˈsiːdɪŋ/ *n.*
action taken in a law court or in a legal case 诉讼

precautionary /prɪˈkɔːʃənəri/ *adj.*
done in order to prevent sth. dangerous or bad from happening 预防的；防范的

controversial /ˌkɒntrəˈvɜːʃəl/ *adj.*
causing a lot of disagreement, because many people have strong opinions about the subject being discussed 引起争论的；有争议的

formulate /ˈfɔːmjʊleɪt/ *vt.*
to develop something such as a plan or set of rules, and decide all the details of how it will be done 构想，制定（计划、规则等）

irreversible /ˌɪrɪˈvɜːsəbl/ *adj.*
impossible to change or bring back a previous condition or situation 不能更改的；不可挽回的；无法逆转的

orverarching /ˌəʊvərˈɑːtʃɪŋ/ *adj.*
affecting or including everything, and therefore very important 支配一切的；包罗万象的；首要的

multitude /ˈmʌltɪtjuːd/ *n.*
an extremely large number of things or people 众多；大量

remediation /rɪˌmiːdɪˈeɪʃən/ *n.*
the process of improving a situation or correcting a problem 补救；纠正

compliance /kəmˈplaɪəns/ *n.*
the practice of obeying rules or requests made by people in authority 服从；顺从；遵从

remedy /ˈremɪdɪ/ *vt.*
to deal with a problem or improve a bad situation 补救；纠正；改善

mandatory /ˈmændətərɪ/ *adj.*
(*formal*) required by law 强制的；法定的；义务的

interrelationship /ˌɪntərɪˈleɪʃənʃɪp/ *n.*
the way in which two or more things or people are connected and affect each other 相互关联；相互影响

collaborative /kəˈlæbəreɪtɪv/ *adj.*
involving, or done by, several people or groups of people working together 合作的；协作的；协力的

stakeholder /ˈsteɪkhəʊldə(r)/ *n.*
a person or company that is involved in a particular organization, project, system, etc., especially because they have invested money in it（某组织、工程、体系等的）参与人，参与方；有权益关系者

vulnerable /ˈvʌlnərəbl/ *adj.*
easily damaged by sth. negative or harmful 易遭毁坏的；易损坏的；脆弱的

Reading comprehension

1 Read Text A and complete the chart to get the outline and main ideas of the text.

Part 1 Principles of environmental law

- 1) _____ may be considered together with integration and interdependence.
- 2) _____ is the right of future generations to enjoy a fair level of the common patrimony.
- 3) _____ is defined as an obligation to protect one's own environment, and to prevent damage to neighboring environments.
- 4) _____ are present in environmental impact assessment, laws.
- 5) _____ was formulated in the Rio Declaration
- 6) _____ is an overarching aim that gives rise to a multitude of legal mechanisms.
- 7) _____ stands for the idea that "the environmental costs of economic activities should be internalized.

Part 2 Approaches of environmental protection

- In some countries, 8) _____ are more commonly used to remedy significant levels of non-compliance with mandatory regulation.
- 9) _____ aims to consider the complex interrelationships of an entire ecosystem in decision-making.
- Some of the most well-known 10) _____ include the Kyoto Protocol and others.

Language focus

1 Match the technical terms in Column A with their Chinese translations in Column B.

Column A	Column B
1. environmental degradation	a. 自愿环境协议
2. intragenerational equity	b. 国际环境协议
3. transboundary responsibility	c. 可持续发展
4. precautionary principle	d. 环境恶化
5. public participation	e. 代内平等
6. sustainable development	f. 跨界责任
7. voluntary environmental agreement	g. 预防原则
8. international environmental agreement	h. 公众参与

2 Fill in the blanks with the words given below. Change the form where necessary.

degrade   mandatory   interdependence   polluter
compliance   ecosystem   assessment   remedy

1. A(n) _____ is all the plants and animals that live in a particular area together with the complex relationship that exists between them and their environment.
2. Our relationship with Earth is one of _____.
3. They can meet the _____ requirement for reducing empty running to cut fuel costs and slashing carbon emissions if companies come together to share the distribution systems.
4. Investigators are sent regularly to verify the _____ with the environmental protection agreement.
5. It is worrying that erosion is _____ the land.
6. A full, highly detailed _____ about what we have on this planet is very important for us human beings.

7. Some say that the _____ should pay the costs of cleaning up the pollution they cause.
8. A great deal has been done internally to _____ the situation.

3 Complete the following sentences by translating the Chinese in brackets into English.

1. People's _____ (获得补偿的权利) may be affected by state and local laws.
2. Since that time, we have been made aware that we may have placed _____ (不合理的经济负担) on those small-sized businesses.
3. The vehicles are not allowed in this city because they are beyond _____ (排放限制).
4. These international environmental agreements have provided the countries concerned with _____ (行为准则).
5. Many governments have taken some steps to _____ (阻止环境恶化).
6. Reaching international agreement on environmental protection will be a long process of _____ (争论和妥协).
7. As part of the _____ (共同的努力) we decided to start an open source project around this core framework of environmental protection.
8. Human overpopulation is _____ (驱动力) behind the current mass extinction crisis and environmental pollution.

4 Translate the following paragraph into English.

中国在公共环境改善、产业升级和发展模式转变方面取得了巨大的进步。中国也已成为国际环境合作的重要力量。中国重视建立一种公平的、合作互利的全球环境治理体系。每个国家都应该依据自己的资源、发展阶段和国内环境，并遵循适合自己发展需要的标准来制定特定而具体的计划。发展中国家应当得到更多的资金、技术和政策支持，以减少他们的温室气体排放，并使其更好地应对气候变化所带来的不可避免的后果。

Critical thinking

1. Some people hold the opinion that environmental protection is the matter of governments at all levels and the general public can only play a minor role in fighting against environmental polution. Work in groups and make comments on this opinion.

2. The problem of environmental pollution should be tackled by a mixture of three elements: technological solutions, laws and regulations, and changes in people's awareness. What would happen if one element disappeared?

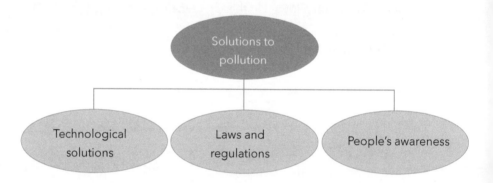

Research task

Academic skill: Making a presentation (Part II)

Start confidently

How you begin your presentation matters a great deal. You will have to gain the audience's attention within the first 10 to 20 seconds. Beginning your presentation with a quick introduction about yourself will help establish your credibility.

Mind your body language

As you begin your presentation, smile and take a deep breath. This will help you relax and dissolve any awkwardness between you and the audience. Mind your posture: Stand straight and hold your head up. Make eye contact with the audience and be mindful of their responses. Feel free to apply any hand gestures, like pointing to the screen or explaining the size of something.

Voice is also important

Make sure that all of your audience can hear you. Talk clearly, loudly, and energetically. But don't be too fast, since there could be people in the audience whose English is not very good. Take advantage of pauses to look up at your audience, give your audience time to react to what you say, or let what you say sink in, or just let yourself breathe and calm down.

Use transitions

Remember to use transitions when moving from one idea to another, like "furthermore", "in addition", "consequently", "meanwhile", "finally", etc. When you talk about the results of something, you can begin with "consequently", or "as a result", etc. When giving a point-by-point explanation, it is better to mention the total number of the points at the outset. For example, "There are reasons for this. The first reason is … The second reason is …", etc. This will help the audience keep track of the points you are talking about. Sometimes a simple pause or a direct statement, such as "Let's move on to the next part of the presentation", is also an effective way to introduce a new part or perspective.

Make sure the closing is natural

Ask if there are any questions, offer your contact information, and tell the audience that you would be open to questions from them over email or phone. Answer the questions, if any; if not, just thank the audience for attending your presentation and walk down the stage.

Task

Collect some information on one approach of environmental protection mentioned in Text A. Work in groups and discuss its feasibility in reality. Then make a presentation with the skills above as well as those discussed in Unit 7.

Section B

Reading strategy

Outlining (Part II)

In Unit 5, we have introduced the steps of outlining and three organizational patterns — order of importance, cause / effect and order of process. In this unit we will introduce another three — time order, classification and comparison / contrast.

Organizational pattern	Explanation	Visual organizer	Signal word
Time order	Details are arranged in the order of occurrence. This is also called chronological order.	Timeline	after, before, during, first, second, third, finally, following, immediately, initially, next, now, preceding, soon, then, today, until, when
Classification	Details are grouped in categories to illustrate or explain a term or concept.	Tree diagram	all, an example of, characterized by, for instance, group, is often called, part of, the other group, category, sort, typically, at the same level
Comparison / Contrast	Details are arranged to show the similarities and differences between and among two or more things.	Venn diagram	although, as well as, but, compared with, different from, however, instead of, like, opposed to, same, similarly, similar to, unlike

The following are examples of the three types of visual organizer.

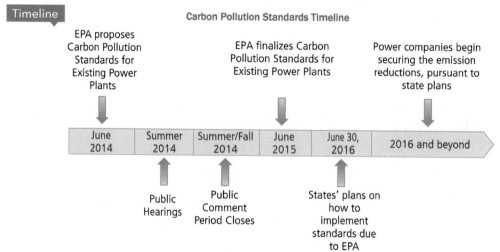

Tree diagram

Venn diagram

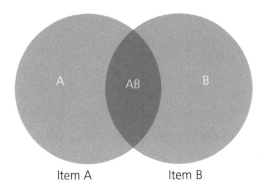

Task

Read Text B and draw an outline to show the causes of the environmental degradation.

Text B

Environmental degradation: causes and effects

1 Environmental degradation is the disintegration of the earth or deterioration of the environment through consumption of assets, for example, air, water and soil; the destruction of environments and the eradication of wildlife. It is characterized as any change or aggravation to nature's turf seen to be pernicious or undesirable. Ecological effect or degradation is created by the

---

deterioration *n.* 恶化
eradication *n.* 根除
aggravation *n.* 加剧；恶化
turf *n.* 地盘；势力范围
pernicious *adj.* 非常有害的；恶性的

consolidation of an effectively substantial and expanding human populace, constantly expanding monetary development or per capita fortune and the application of asset exhausting and polluting technology. It occurs when Earth's natural resources are depleted and environment is compromised in the form of extinction of species, pollution in air, water and soil, and rapid growth in population.

2 Environmental degradation is one of the largest threats that are being looked at in the world today. The United Nations International Strategy for Disaster Reduction characterizes environmental degradation as the lessening of the limit of the Earth to meet social and environmental destinations and needs. Environmental degradation can happen in a number of ways and the causes of it are varied. At the point when environments are wrecked or common assets are exhausted, the environment is considered to be corrupted and harmed.

3 **Land disturbance.** A more basic cause of environmental degradation is land disturbance. Numerous weedy plant species, for example, garlic mustard, are both foreign and obtrusive. A rupture in the environmental surroundings provides for them a chance to start growing and spreading. These plants can assume control over nature, eliminating the local greenery. The result is territory with a solitary predominant plant which doesn't give satisfactory food assets to all the environmental life. The whole environment can be destroyed because of these invasive species.

4 **Pollution.** Pollution, in whatever form, whether it is air, water, land or noise, is harmful for the environment. Air pollution pollutes the air that we breathe, which causes health issues. Water pollution degrades the quality of water that

deplete *vt.* 使耗尽；使枯竭
wrecked *vt.* 使（某物）严重损毁
obtrusive *adj.* 突兀的
rupture *n.* 破裂
invasive *adj.* 入侵的，侵害的

we use for drinking purposes. Land pollution results in degradation of Earth's surface as a result of human activities. Noise pollution can cause damage to our ears when exposed to continuous large sounds like honking of vehicles on a busy road or machines producing large noise in a factory or a mill.

5 **Overpopulation.** Rapid population growth puts strain on natural resources which results in degradation of our environment. Mortality rate has gone down due to better medical facilities which have resulted in increased lifespan. More population simply means more demand for food, clothes and shelter. You need more space to grow food and provide homes to millions of people. This results in deforestation, which is another factor of environmental degradation.

6 **Landfills.** Landfills pollute the environment and destroy the beauty of the city. Landfills come within the city due to the large amount of waste generated by households, industries, factories and hospitals. Landfills pose a great risk to the health of the environment and the people who live there. Landfills produce foul smell when burned and cause huge environmental degradation.

7 **Deforestation.** Deforestation is the cutting down of trees to make way for more homes and industries. Rapid growth in population and urban sprawl are two of the major causes of deforestation. Apart from that, use of forest land for agriculture, animal grazing, harvest for fuel wood and logging are some of the other causes of deforestation. Deforestation contributes to global warming as decreased forest size puts carbon back into the environment.

8 **Natural causes.** Things like avalanches, quakes, tidal waves, storms, and wildfires can totally crush nearby animal and plant groups to the point where they can no longer survive in those areas. This can either come to fruition through physical demolition as the result of a specific disaster, or by the long-

honk *vi.* 鸣喇叭
mortality rate 死亡率
avalanche *n.* 雪崩
demolition *n.* 拆除

term degradation of assets by the presentation of an obtrusive foreign species to the environment. The latter frequently happens after tidal waves, when reptiles and bugs are washed ashore.

9 **Earth itself.** Of course, humans aren't totally to blame for this whole thing. Earth itself causes ecological issues as well. While environmental degradation is most normally connected with the things that people do, the truth of the matter is that the environment is always changing. With or without the effect of human exercises, a few biological systems degrade to the point where they can't help the life that is supposed to live there.

10 Human health might be at the receiving end as a result of the environmental degradation. Areas exposed to toxic air pollutants can cause respiratory problems like pneumonia and asthma. Millions of people are known to have died due to indirect effects of air pollution.

11 Some effects of environmental degradation are found in different areas.

12 **Biodiversity.** Biodiversity is important for maintaining balance of the ecosystem in the form of combating pollution, restoring nutrients, protecting water sources and stabilizing climate. Deforestation, global warming, overpopulation and pollution are few of the major causes for loss of biodiversity.

13 **Ozone layer.** Ozone layer is responsible for protecting earth from harmful ultraviolet rays. The presence of chlorofluorocarbons, hydro chlorofluorocarbons in the atmosphere is causing the ozone layer to deplete. As it will deplete, it will emit harmful radiations back to the earth.

---

obtrusive *adj.* 突出的；显眼的
respiratory *adj.* 呼吸的
pneumonia *n.* 肺炎
asthma *n.* 哮喘
chlorofluorocarbon *n.* 氯氟烃

14. **Deterioration.** The deterioration of environment can be a huge setback for tourism industry that rely on tourists for their daily livelihood. Environmental damage in the form of loss of green cover, loss of biodiversity, huge landfills, increased air and water pollution can be a big turn-off for most of the tourists.

15. **Huge cost.** The huge cost that a country may have to bear due to environmental degradation can have big economic impact in terms of restoration of green cover, cleaning up of landfills and protection of endangered species. The economic impact can also be in terms of loss of tourism industry.

16. As you can see, there are a lot of things that can have an effect on the environment. If we are not careful, we can worsen environmental degradation. We can, however, take action to stop it and take care of the world that we live in by providing environmental education to the people to help them learn how to take care of the environment, thus making our home planet better protected for future generations.